T0265351

Second Edition

FUNCTIONAL ANALYSIS

Entering Hilbert Space

Vagn Lundsgaard Hansen

Technical University of Denmark, Denmark

Second Edition

FUNCTIONAL ANALYSIS

Entering Hilbert Space

World Scientific

NEW JERSEY · LONDON · SINGAPORE · BEIJING · SHANGHAI · HONG KONG · TAIPEI · CHENNAI · TOKYO

Published by

World Scientific Publishing Co. Pte. Ltd.

5 Toh Tuck Link, Singapore 596224

USA office: 27 Warren Street, Suite 401-402, Hackensack, NJ 07601

UK office: 57 Shelton Street, Covent Garden, London WC2H 9HE

Library of Congress Cataloging-in-Publication Data
Names: Hansen, Vagn Lundsgaard.
Title: Functional analysis : entering Hilbert space / by Vagn Lundsgaard Hansen,
    Technical University of Denmark, Denmark.
Description: 2nd edition. | New Jersey : World Scientific, 2016. |
    Includes bibliographical references and index.
Identifiers: LCCN 2015038647 | ISBN 9789814733922 (hardcover : alk. paper)
Subjects: LCSH: Functional analysis. | Hilbert space.
Classification: LCC QA320 .H36 2016 | DDC 515/.733--dc23
LC record available at http://lccn.loc.gov/2015038647

British Library Cataloguing-in-Publication Data
A catalogue record for this book is available from the British Library.

Printed in Singapore

To my students

Preface to the Second Edition

In the second edition, I have expanded the material on normed vector spaces and their operators presented in Chapter 1 to include proofs of the Open Mapping Theorem, the Closed Graph Theorem and the Hahn-Banach Theorem.

The material on operators between normed vector spaces is further expanded in a new chapter on Fredholm theory (Chapter 6). Fredholm theory originates in pioneering work of the Swedish mathematician Erik Ivar Fredholm on integral equations, which inspired the study of a new class of bounded linear operators, known as Fredholm operators. Chapter 6 presents the basic elements of the theory of Fredholm operators on general Banach spaces, not only on Hilbert spaces, since this is important for applications of the theory. The more general setting with Banach spaces requires that we develop the theory of dual operators between Banach spaces to replace the use of adjoint operators between Hilbert spaces.

Fredholm operators are of interest far beyond mathematical analysis, they also play a significant role in theoretical physics, differential geometry and topology with the famous Index Theorem proved by Michael Atiyah and Isadore Singer in 1963 as a highlight.

With the addition of the new material on normed vector spaces and their operators, the book can hopefully serve as a general introduction to functional analysis viewed as a theory of infinite dimensional linear spaces and linear operators acting on them.

Technical University of Denmark, 2015

Vagn Lundsgaard Hansen

Preface to the First Edition

Functional analysis is one of the important new mathematical fields from the twentieth century. It originates from the calculus of variations in the study of operators on function spaces defined by differentiation and integration. The name functional analysis was coined by the French mathematician Paul Lévy (1886–1971). Early pioneers were Italian Vito Volterra (1860–1940), Swedish Erik Ivar Fredholm (1866–1927), German David Hilbert (1862–1943) and Hungarian Frigyes Riesz (1880–1956). The Polish mathematician Stefan Banach (1892–1945) was influential in bringing the notions of topology into functional analysis, and he is known for the seminal book *Théorie des opérations linéaires* of 1932. Fundamental contributions to the study of operators on Hilbert spaces were made by the Hungarian mathematician John von Neumann (1903–57). Inspiration to this work came from many sides, not least from the development of quantum mechanics in physics in the 1920s by physicists such as Niels Bohr (1885–1962), Paul Dirac (1902–84), Werner K. Heisenberg (1901–76) and Erwin Schrödinger (1887–1961).

The present book supplements my earlier book *Fundamental Concepts in Modern Analysis*, World Scientific, 1999, with an introduction to Hilbert spaces. The new book can be read independently of the first book by readers with a basic knowledge in general topology and normed vector spaces, including the operator norm of a bounded linear operator. To set the stage properly we review briefly the necessary elements of metric topology in Chapter 1. Together the two books are used for a course at the advanced undergraduate/beginning graduate level offered to engineering students at the Technical University of Denmark. It is my hope that the books will also appeal to university students in mathematics and in the physical sciences.

A fundamental problem in applications of functional analysis is connected with the construction of suitable spaces to handle a given application. Often the spaces needed are complete normed vector spaces (Banach spaces) constructed from spaces of continuous functions. In this spirit, Chapter 2 opens with a slightly novel construction of the L^p-spaces together with proofs of the basic inequalities of Hölder and Minkowski in these spaces. Other important Banach spaces arising from function spaces and sequence spaces are also treated.

The main bulk of the book presents the basic elements of the theory of Hilbert spaces (Chapter 3) and operators on Hilbert spaces (Chapter 4), culminating in a proof of the spectral theorem for compact, self-adjoint operators on separable Hilbert spaces (Chapter 5).

The presentation of material has been greatly influenced by ideas of my good colleague Michael Pedersen presented in his book *Functional Analysis in Applied Mathematics and Engineering*, Chapman & Hall/CRC, 2000. I appreciate many joyful conversations with Michael about functional analysis and thank him for several valuable comments to my exposition. I also like to point out that his book contains a wealth of applications of methods from functional analysis in the engineering sciences, in particular applications based on the abstract formulations of partial differential equations in a Hilbert space setting.

I am extremely grateful to my colleagues Preben Alsholm and Ole Jørsboe for a very careful reading of the entire manuscript and for many valuable suggestions.

Notions from general topology and basic elements of the theory of normed vector spaces are in general used throughout this book without further notice. To ease the reading of the book, a short summary of the necessary elements of metric topology is given in Chapter 1.

Technical University of Denmark, 2006

Vagn Lundsgaard Hansen

Preliminary Notions

There are certain standard notations and terminologies used throughout mathematics. In this explanatory note we list some of these.

Logical symbols

$\forall$	for all
$\exists$	there exists
$\implies$	implies
$\iff$	if and only if
$\in$	belongs to
$\notin$	does not belong to
$\subset$	proper subset
$\subseteq$	subset
$\wedge$	and
$\vee$	or
$:$	such that

Sets of numbers

$\mathbb{N}$	the set of natural numbers
$\mathbb{Z}$	the set of integers
$\mathbb{Q}$	the set of rational numbers
$\mathbb{I}$	the set of irrational numbers
$\mathbb{R}$	the set of real numbers
$\mathbb{R}^+$	the set of positive real numbers
$\mathbb{R}_0^+$	the set of non-negative real numbers
$\mathbb{C}$	the set of complex numbers

Notions from set theory

A set can be declared by listing the elements. For example,

$$\{x \mid x \text{ has property } \mathcal{P}\}$$

denotes the set of elements x with a given property $\mathcal{P}$. Often the property $\mathcal{P}$ will be expressed in logical symbols as below.

$\emptyset$ the empty set

$A \cup B$ the union of sets A and B, i.e. $\{x \mid x \in A \vee x \in B\}$

$A \cap B$ the intersection of sets A and B, i.e. $\{x \mid x \in A \wedge x \in B\}$

$A \sqcup B$ the union of disjoint sets A and B,
         i.e. $\{x \mid x \in A \vee x \in B\}$ and $A \cap B = \emptyset$

$\displaystyle\bigcup_{\alpha \in I} A_\alpha$ the union of sets A_α indexed by
         the elements α in an index set I

$\displaystyle\bigcap_{\alpha \in I} A_\alpha$ the intersection of sets A_α indexed by
         the elements α in an index set I

$A \times B$ the product set of the sets A and B,
         i.e. $\{(x, y) \mid x \in A, \ y \in B\}$

$S \setminus A$ the set difference, i.e. $\{x \in S \mid x \notin A\}$

Notions related to mappings

$f : A \to B$ a mapping of A into B

$f(A)$ the *image* of f, i.e. $f(A) = \{b \in B \mid \exists a \in A : f(a) = b\}$

$f^{-1}(C)$ the *preimage* of subset $C \subseteq B$ under mapping $f : A \to B$,
         i.e. $f^{-1}(C) = \{a \in A \mid f(a) \in C\}$

Relations and equivalence classes

A *relation* $\sim$ in a set S is a subset $R \subseteq S \times S$, or in other words, a distinguished set of ordered pairs of points $x, y \in S$. We write $x \sim y$ if $(x, y) \in R$.

The relation $\sim$ is called an *equivalence relation* if

(i) $x \sim x$ ($\sim$ is reflexive)

(ii) $x \sim y \implies y \sim x$ ($\sim$ is symmetric)

(iii) $(x \sim y) \wedge (y \sim z) \implies x \sim z$ ($\sim$ is transitive) .

If $\sim$ is an equivalence relation in S, then S can be partitioned into a corresponding system of disjoint subsets, so-called *equivalence classes* S_α, indexed by $\alpha \in I$, and defined by

$$x, y \in S_\alpha \iff x \sim y .$$

If on the other hand,

$$S = \bigsqcup_{\alpha \in I} S_\alpha ,$$

then we can define a relation $\sim$ in S by

$$x \sim y \iff x, y \in S_\alpha \text{ for some } \alpha \in I .$$

Clearly $\sim$ is an equivalence relation in S.

An equivalence relation in a set S and a partition of S into a disjoint union of equivalence classes amounts in other words to the same thing.

If $\sim$ is an equivalence relation in S, the set of equivalence classes is denoted by $\tilde{S} = S/\sim$ and $\pi : S \to \tilde{S}$ denotes the mapping which associates to an element $x \in S$ its equivalence class $\pi(x) \in \tilde{S}$.

Notions related to orderings

A (partial) *ordering* $\leq$ in a set S is a relation $R \subseteq S \times S$, where we write $x \leq y$ if $(x, y) \in R$, that satisfies the following conditions:

(i) $x \leq x$ ($\leq$ is reflexive)

(ii) $(x \leq y) \wedge (y \leq x) \implies x = y$ ($\leq$ is anti-symmetric)

(iii) $(x \leq y) \wedge (y \leq z) \implies x \leq z$ ($\leq$ is transitive).

A set S together with an ordering relation is called an *ordered set*. In this book, the only ordering relation we shall consider is the usual ordering $\leq$ in the set of real numbers $\mathbb{R}$.

Let A be a subset in the ordered set S with ordering relation $\leq$.

An element $x_{\max} \in A$ is called a *maximal element* in A if $x \leq x_{\max}$ for all $x \in A$. Similarly, an element $x_{\min} \in A$ is called a *minimal element* in A if $x_{\min} \leq x$ for all $x \in A$.

An element $s \in S$ is called an *upper bound* for A if $x \leq s$ for all $x \in A$. If A is *bounded from above*, then an upper bound $s^* \in S$ is called a *least upper bound*, or a *supremum*, of A, if $s^* \leq s$ for all other upper bounds $s \in S$ of A.

An element $t \in S$ is called a *lower bound* for A if $t \leq x$ for all $x \in A$. If A is *bounded from below*, then a lower bound $t^* \in S$ is called a *greatest lower bound*, or an *infimum*, of A, if $t \leq t^*$ for all other lower bounds $t \in S$ of A.

Contents

Chapter 1

Basic Elements of Metric Topology

A major step forward in the development of functional analysis was taken around 1900 when mathematicians in their search for solutions to variational problems began to consider functions with specified properties as elements in spaces of functions of similar type. This was enhanced by the creation of new mathematical structures, such as metric spaces and more generally, topological spaces, which came to be known as general, or pointset, topology. Using these new structures a firm basis could be established for fundamental notions from analysis such as convergence of sequences and continuity of functions.

In this chapter we review briefly necessary basic elements of general topology of spaces involving a metric structure. We develop basic material on normed vector spaces and their operators with full proofs of the Open Mapping Theorem, the Closed Graph Theorem and the Hahn-Banach Theorem.

1.1 Metric spaces

A distance function (or a metric) on a set M is a function $d : M \times M \to \mathbb{R}$ which to any pair of points $x, y \in M$ associates a real number $d(x, y)$, called the *distance* from x to y. Furthermore, to get a reasonable notion of distance, it has proved fruitful to require that the following three conditions are satisfied:

MET 1 (positive definite)
$$d(x, y) \geq 0, \text{ for all } x, y \in M, \quad = 0 \iff x = y.$$

MET 2 (symmetry)
$$d(x, y) = d(y, x), \text{ for all } x, y \in M.$$

MET 3 (the triangle inequality)
$$d(x, z) \leq d(x, y) + d(y, z), \text{ for all } x, y, z \in M.$$

Definition 1.1.1. A *metric* on a set M is a function $d : M \times M \to \mathbb{R}$ satisfying the conditions MET 1, MET 2, MET 3.

A pair (M, d) consisting of a set M together with a specific metric d on M is called a *metric space.*

Remark 1.1.2. The same set M can be equipped with several different metrics, in which case the corresponding metric spaces are counted as different.

When the context leaves no doubt as to which metric is being considered, usually the metric is not mentioned explicitly.

The idea of metric spaces was introduced by the French mathematician Maurice Fréchet (1878–1973) in his doctoral thesis of 1906.

A few examples of metric spaces are in order.

Example 1.1.3. Consider the set of real n-tuples $M = \mathbb{R}^n$.

For points $x = (x_1, \ldots, x_n)$ and $y = (y_1, \ldots, y_n)$ in $\mathbb{R}^n$ we set

$$d(x, y) = \sqrt{\sum_{i=1}^{n} (x_i - y_i)^2} \ .$$

Clearly MET 1 and MET 2 are satisfied. To prove that MET 3 is satisfied we need some preparations.

The space $\mathbb{R}^n$ is an n-dimensional real vector space with the usual coordinatewise definition of addition and multiplication by scalars. The points in $\mathbb{R}^n$ are then identified with vectors. We define the *inner product* by

$$(x, y) = \sum_{i=1}^{n} x_i y_i \ ,$$

and the associated *norm* by

$$\|x\| = \sqrt{\sum_{i=1}^{n} x_i{}^2} \ .$$

Then we have the *Cauchy-Schwarz inequality:*

$$|(x, y)| \le \|x\| \|y\| \quad \text{for all} \quad x, y \in \mathbb{R}^n \ .$$

Proof. Let $x, y \in \mathbb{R}^n$ be arbitrarily chosen, but fixed vectors in $\mathbb{R}^n$. For every real number $t \in \mathbb{R}$ we have the inequality

$$0 \le (x + ty, x + ty) = \|y\|^2 t^2 + 2(x, y)t + \|x\|^2 \ .$$

For $y \neq 0$, this describes a polynomial of degree 2 in t (a parabola) with at most one zero. Therefore, the discriminant of the polynomial must satisfy

$$4(x, y)^2 - 4||x||^2||y||^2 \leq 0 .$$

From this we get the inequality $|(x, y)| \leq ||x|| \, ||y||$ as asserted. For $y = 0$, the inequality is trivially satisfied. □

The Cauchy-Schwarz inequality implies the *triangle inequality:*

$$||x + y|| \leq ||x|| + ||y||.$$

Proof. Using the Cauchy-Schwarz inequality, the small computation

$$
\begin{aligned}
||x + y||^2 &= (x + y, x + y) = ||x||^2 + ||y||^2 + 2(x, y) \\
&\leq ||x||^2 + ||y||^2 + 2||x|| \cdot ||y|| \\
&= (||x|| + ||y||)^2
\end{aligned}
$$

reveals the triangle inequality. □

By noting that

$$d(x, y) = ||x - y||,$$

the condition MET 3 now follows immediately:

$$
\begin{aligned}
d(x, z) = ||x - z|| &= ||x - y + y - z|| \\
&\leq ||x - y|| + ||y - z|| = d(x, y) + d(y, z).
\end{aligned}
$$

Thereby we have verified that $\mathbb{R}^n$ equipped with the distance function d is a metric space. This metric space is called n-dimensional *Euclidean space*, and the metric d, *the Euclidean metric* on $\mathbb{R}^n$.

For $n = 1$ we get the usual concept of distance on a line by identification with $\mathbb{R}$ along a coordinate axis. For $n = 2, 3$ we rediscover the usual concept of distance in a plane, respectively 3-dimensional space, when these are identified with $\mathbb{R}^2$, respectively $\mathbb{R}^3$, by choosing an orthogonal coordinate system.

Before the next example, we recall a fundamental property of the system of real numbers, namely the existence of a supremum (and infimum). Together with the well-known algebraic properties, the existence of a supremum characterizes the system of real numbers.

Property (Supremum). *Every non-empty subset A of $\mathbb{R}$ which is bounded from above has a smallest upper bound, which is called the* least *upper bound of A, or supremum of A, and is denoted by* sup A.

From the existence of a supremum we easily get:

Property (Infimum). *Every non-empty subset A of $\mathbb{R}$ which is bounded from below has a largest lower bound, which is called the* greatest lower bound *of A, or* infimum *of A, and is denoted by* $\inf A$.

Example 1.1.4. For K a non-empty, but otherwise arbitrary, fixed set, let M be the set of all bounded real-valued functions $f : K \to \mathbb{R}$ defined on K.
For $f, g \in M$ put

$$d(f,g) = \sup_{x \in K} |f(x) - g(x)|$$
$$= \sup \left\{ |f(x) - g(x)| \mid x \in K \right\}.$$

Note that since

$$A = \left\{ |f(x) - g(x)| \mid x \in K \right\}$$

is bounded from above, the number $d(f,g) = \sup A$ exists. Note also that we have used different ways of writing $\sup A$ in the present example.

MET 1 follows since $0 \le |f(x) - g(x)| \le \sup A$ for all $x \in K$. MET 2 is also easily seen, since $|f(x) - g(x)| = |g(x) - f(x)|$ for all $x \in K$.

To prove that MET 3 is satisfied we proceed as follows.

Let f, g, h be three functions in M. For every $x \in K$, we then have

$$|f(x) - h(x)| = |f(x) - g(x) + g(x) - h(x)|$$
$$\le |f(x) - g(x)| + |g(x) - h(x)|$$
$$\le \sup_{x \in K} |f(x) - g(x)| + \sup_{x \in K} |g(x) - h(x)|$$
$$= d(f,g) + d(y,h).$$

From this we see that $d(f,g) + d(g,h)$ is an upper bound for

$$\left\{ |f(x) - h(x)| \mid x \in K \right\}.$$

Since any upper bound is greater than or equal to the least upper bound, we conclude that

$$d(f,h) = \sup_{x \in K} |f(x) - h(x)| \le d(f,g) + d(g,h).$$

Hence d is a metric on M.

If we define addition of functions in M and multiplication of functions by real numbers using the obvious pointwise defined operations, M gets the structure of a real vector space. This vector space has finite dimension if K

contains only a finite number of points. If K contains infinitely many points there will be no system of finitely many functions spanning M and hence the vector space M is infinite dimensional in this case.

Example 1.1.5. Let M be any set. Set

$$d(x, y) = \begin{cases} 1 \text{ for } x \neq y \\ 0 \text{ for } x = y. \end{cases}$$

It is easy to see that d is a metric on M. This metric on M is called *the discrete metric*.

Example 1.1.6. Let (M, d) be a metric space, and let $A \subseteq M$ be a subset of M. Then A inherits a metric from (M, d) called *the induced metric*. More formally: If $d : M \times M \to \mathbb{R}$ is the metric on M, we get the induced metric on A by taking the restriction of d to $A \times A$.

We finish this section with the definition of the notion of continuity of a mapping in the setting of metric spaces.

Definition 1.1.7. Let (X, d_X) and (Y, d_Y) be metric spaces. We say that the mapping $f : X \to Y$ is *continuous at a point* $x_0 \in X$, if

$$\forall \varepsilon > 0 \ \exists \delta > 0 \ \forall x \in X : d_X(x, x_0) < \delta \quad \Rightarrow \quad d_Y(f(x), f(x_0)) < \varepsilon.$$

We say that the mapping $f : X \to Y$ is *continuous* if f is continuous at every point $x_0 \in X$.

1.2 The topology of a metric space

A closer study of continuity of mappings in the setting of metric spaces reveals that it is not the specific metrics as such that are decisive for continuity to make sense but rather a class of subsets defined by the metrics leading to the concept of the underlying *topology* in a metric space.

First some definitions. Let (M, d) be an arbitrary metric space, let x_0 be a point in M, and let $r \in \mathbb{R}^+$ be a positive real number. Then the set

$$B_r(x_0) = \{x \in S \mid d(x_0, x) < r\}$$

is called the *open ball* or the *open sphere* in M with *center* x_0 and *radius* r.

Definition 1.2.1. Let (M, d) be a metric space. A subset W of M is called an *open set* in the metric space (M, d), if for every $x \in W$ there exists a $\delta > 0$ such that $B_\delta(x) \subseteq W$.

Remark 1.2.2. An open ball $B_r(x_0)$ in the metric space (M, d) is an open set in M, since for $x \in B_r(x_0)$ the triangle inequality shows that $B_\delta(x) \subseteq B_r(x_0)$, if $0 < \delta \le r - d(x_0, x)$.

Continuity of a mapping can be described in a very simple way using open sets. The proof of the following theorem is left to the reader.

Theorem 1.2.3. *Let $f : X \to Y$ be a mapping between metric spaces (X, d_X) and (Y, d_Y). Then f is continuous if and only if for every open set V in Y, the set $f^{-1}(V)$ is an open set in X.*

Theorem 1.2.3 shows that the family of open sets in a metric space plays a decisive role in the study of continuity of mappings between metric spaces. The main features of the family of open sets are captured by three fundamental properties, the proofs of which are left to the reader.

Theorem 1.2.4. *The family of open sets $\mathcal{T}$ in a metric space (M, d) has the following fundamental properties.*

TOP 1 *If $\{U_i \in \mathcal{T} \mid i \in I\}$ is an arbitrary system of subsets in M from $\mathcal{T}$, then the union $\cup\{U_i \in \mathcal{T} \mid i \in I\}$ of these subsets also belongs to $\mathcal{T}$.*

TOP 2 *If $U_1, \ldots, U_k$ is an arbitrary finite system of subsets in M from $\mathcal{T}$, then the intersection $\cap_{i=1}^{k} U_i$ of these subsets also belongs to $\mathcal{T}$.*

TOP 3 *The empty set $\emptyset$, and the set M itself belong to $\mathcal{T}$.*

A family of subsets $\mathcal{T}$ in an arbitrary set M with the properties TOP 1, TOP 2, TOP 3 is called a *topology* in M. In this general context, the pair $(M, \mathcal{T})$ is called a *topological space* and the subsets of M belonging to the family $\mathcal{T}$ is referred to as the family of open sets in the topological space. The family of open sets in a metric space is called the topology in the metric space.

Complementary to the family of open sets we have the family of closed sets.

Definition 1.2.5. A subset A of a metric space (M, d), or more generally a topological space $(M, \mathcal{T})$, is said to be *closed* if the complementary set $M \setminus A$ of A is an open set in M.

The main reason for introducing the family of closed sets in a topological space is that it is easier to use in certain situations. The family of closed sets has properties complementary to the properties TOP 1, TOP 2, TOP 3, and continuity of mappings can be expressed using closed sets in complete analogy with using open sets.

For an arbitrary subset W in a metric space (M, d), or more generally a topological space $(M, \mathcal{T})$, there is an *interior* int$W$ and a *closure* $\overline{W}$, which is respectively the largest open set contained in W and the smallest closed set containing W.

More precisely, the interior int$W$ of a subset W in a metric space (M, d) consists of all the points $x \in W$ for which sufficiently small open balls centered at x are completely contained in W, so-called *interior points*. Similarly, the closure $\overline{W}$ consists of all points $x \in S$ for which every ball centered at x contains points from W, so-called *contact points*.

The important property of a subset being dense in a metric space can be expressed using the closure operation.

Definition 1.2.6. A subset W of a metric space (M, d), or more generally a topological space $(M, \mathcal{T})$, is said to be *dense* in M if all points in M are contact points for W, in other words if the closure of W satisfies $\overline{W} = M$.

1.3 Completeness and compactness of metric spaces

This section is devoted to some fundamental notions associated with sequences in a metric space M with metric d.

Definition 1.3.1. Let (x_n), or more explicitly $x_1, x_2, \ldots, x_n, \ldots$, be a sequence of points in the metric space (M, d), and let $y_0 \in M$. We say that

$$x_n \text{ has the } \textit{limit point } y_0 \text{ for } n \textit{ going to } \infty ,$$

or that

$$x_n \textit{ converges to } y_0 \text{ for } n \textit{ going to } \infty ,$$

if

$$\forall \varepsilon > 0 \ \exists n_0 \in \mathbb{N} \ \forall n \in \mathbb{N} : \ n \geq n_0 \quad \Rightarrow \quad d(x_n, y_0) < \varepsilon .$$

A sequence (x_n) in a metric space (M, d) has at most one limit point. This is easily seen by using the triangle inequality.

The French mathematician Augustin-Louis Cauchy (1789–1857), who is one of the pioneers in mathematical analysis, has formulated a convergence principle for sequences of real numbers, which has inspired the following definition.

Definition 1.3.2. A sequence (x_n) of points in (M, d) is called a *Cauchy sequence*, or a *fundamental sequence*, if for every $\varepsilon > 0$, there exists an $n_0 \in \mathbb{N}$ such that $d(x_n, x_m) < \varepsilon$ for all $n, m \geq n_0$. Or, using quantifiers:

$$\forall \varepsilon > 0 \ \exists n_0 \in \mathbb{N} \ \forall n, m \in \mathbb{N}: \qquad n, m \geq n_0 \quad \Rightarrow \quad d(x_n, x_m) < \varepsilon.$$

An equivalent, though in certain contexts more convenient, formulation of the concept of a Cauchy sequence can be expressed as follows using quantifiers:

$$\forall \varepsilon > 0 \ \exists n_0 \in \mathbb{N} \ \forall n, k \in \mathbb{N}: \qquad n \geq n_0 \quad \Rightarrow \quad d(x_n, x_{n+k}) < \varepsilon.$$

It is easy to prove that every convergent sequence (x_n) in a metric space (M, d) is a Cauchy sequence. But the converse is not true. It depends on the metric space, whether *all* Cauchy sequences in the space are convergent.

Example 1.3.3. Consider an arbitrary metric space (M', d') containing a convergent sequence (x_n) of pairwise different elements with limit point y_0. Remove the point y_0 from M' to form the subspace $M = M' \setminus \{y_0\}$. Equip M with the metric d induced from d'. Then the sequence (x_n) is a Cauchy sequence in the metric space (M, d), which is not convergent to a point in M.

Metric spaces in which all Cauchy sequences are convergent are, however, sufficiently abundant and interesting that we introduce the following definition.

Definition 1.3.4. A metric space (M, d) is said to be *complete* if every Cauchy sequence in (M, d) is convergent.

An important result from classical analysis can then be expressed as follows.

Theorem 1.3.5 (The Cauchy condition for sequences). *The set $\mathbb{R}$ of real numbers with the usual metric is a complete metric space.*

Although Example 1.3.3 may seem artificial, it is the typical way that a metric space can fail to be complete. We shall return to this in Theorem 2.1.6.

It is easy to prove that completeness is inherited by closed subsets.

Theorem 1.3.6. *Let (M, d) be a complete metric space, and let A be a closed subset in (M, d). Then A with the metric induced from (M, d) is a complete metric space.*

Another important concept in topology is the notion of *compactness*. In functional analysis, where most spaces considered carry metrics, compactness is normally introduced using sequences.

Definition 1.3.7. A subset K in a metric space (M, d) is said to be *sequentially compact*, or just *compact*, if every sequence (x_n) of elements in K contains a convergent subsequence (x_{n_k}) with limit point $x \in K$.

Remark 1.3.8. The whole basic set M in a metric space (M, d) can, of course, in itself be compact.

Definition 1.3.9. A subset K in a metric space (M, d) is said to be *bounded* if it is contained in an open ball, i.e. if there exist a point $x \in M$ and a radius $r > 0$ such that $K \subseteq B_r(x)$.

There are many important connections between the concepts just introduced. The following two results are not particularly difficult to prove.

Theorem 1.3.10. *Every compact metric space (M, d) is complete.*

Theorem 1.3.11. *Every compact subset K in a metric space (M, d) is closed and bounded.*

In $\mathbb{R}^n$ with the Euclidean metric, the converse to Theorem 1.3.11 is also true. This is the main content of the famous Heine-Borel Theorem, which we state without proof.

Theorem 1.3.12 (Heine-Borel). *A subset K in $\mathbb{R}^n$ is compact if and only if it is closed and bounded.*

In infinite dimensional function spaces, compactness is more subtle than in finite dimensional spaces.

For a family of functions to form a compact subset in a function space, we can, however, formulate a useful sufficient condition in the spirit of Theorem 1.3.12 in which closed subsets of functions are substituted with equicontinuous subsets of functions. Loosely speaking, a family of mappings between metric spaces is called equicontinuous, if each map in the family is uniformly continuous and if for every given tolerance $\varepsilon > 0$ in the output, we can choose the same tolerance $\delta > 0$ in the input for all maps in the family. Expressed formally using quantifies, the definition is as follows.

Definition 1.3.13. Let (X, d_X) and (Y, d_Y) be metric spaces, and let $\mathcal{E}$ be an arbitrary family of maps $f : X \to Y$. Then $\mathcal{E}$ is said to be *equicontinuous* if

$$\forall \varepsilon > 0 \ \exists \delta > 0 \ \forall f \in \mathcal{E} \ \forall x, y \in X : d_X(x, y) < \delta \quad \Rightarrow \quad d_Y(f(x), f(y)) < \varepsilon.$$

Theorem 1.3.14 (Ascoli-Arzela Theorem). *Let K be a compact subset in a metric space (M, d) and let $C(K)$ denote the space of continuous, complex-valued functions $f : K \to \mathbb{C}$, equipped with the uniform metric $D(f, g) = \sup_{x \in K} |f(x) - g(x)|$. Then a subset $\mathcal{K}$ in $(C(K), D)$ is compact if it is equicontinuous and bounded.*

Proof. Suppose $\mathcal{K}$ is an equicontinuous and bounded subset in $(C(K), D)$, and let (f_n) be an arbitrary sequence in $\mathcal{K}$.

Since (f_n) is bounded in the uniform metric, the closure of the set of elements in the sequence $(f_n(x))$ is a closed and bounded subset of complex numbers for each $x \in K$, and hence a compact subset of $\mathbb{C}$ by Theorem 1.3.12. Consequently the sequence $(f_n(x))$ contains a convergent subsequence for each $x \in K$.

Since K is compact, we can for any given radius $r > 0$, choose a finite set of points $x_1, \ldots, x_k \in K$, such that $K \subseteq B_r(x_1) \cup \cdots \cup B_r(x_k)$. This follows easily since the existence of an infinite sequence (x_n) of elements in K for which $d(x_i, x_j) \geq r$ for $i \neq j$ would contradict that K is compact.

For each $m \in \mathbb{N}$, we can therefore choose a finite set of points

$$W_m = \{x_{m1}, \ldots, x_{mj_m}\} \subseteq K \ ,$$

such that $K \subseteq B_{1/m}(x_{m1}) \cup \cdots \cup B_{1/m}(x_{mj_m})$.

By the construction, it is clear that the union $W = \cup_{m=1}^{\infty} W_m$ of the subsets W_m is a countable dense subset in K, i.e. $K = \overline{W}$.

Let $I = \{(m, j) \, | \, m \in \mathbb{N}, j = 1, \ldots, j_m\}$ be the index set for the points in W equipped with the lexicographical ordering relation, and let $\tau : \mathbb{N} \to I$ be the unique bijective, order preserving (counting) map for I with $\tau(1) = (1, 1)$.

For $k = 1$ we can choose a subsequence (f_n^1) of (f_n) for which the sequence of complex numbers $(f_n^1(x_{\tau(1)}))$ is a convergent subsequence of the sequence $(f_n(x_{\tau(1)}))$. By induction we can then successively choose a subsequence (f_n^k) of (f_n^{k-1}) for each $k > 1$, where the sequence of complex numbers $(f_n^k(x_{\tau(k)}))$ is a convergent subsequence of the sequence $(f_n(x_{\tau(k)}))$. In this way we obtain that the subsequences $(f_n^k(x_{mj}))$ of $(f_n(x_{mj}))$ converges at any point $x_{mj} \in W$ for k sufficiently large, since convergence of a sequence only depends on the 'tail' of the sequence.

Now let (f_n^∞) be the subsequence of (f_n) in which the kth function is the kth function in the subsequence (f_n^k), i.e. (f_n^∞) is the diagonal sequence of the subsequences (f_n^k). By the construction it follows that the sequence of complex numbers $(f_n^\infty(x_{mj}))$ converges at any point $x_{mj} \in W$.

We shall now prove that the subsequence (f_n^∞) of the original sequence (f_n)

converges in $(C(K), D)$. It is easy to prove that $(C(K), D)$ is a complete metric space by substituting the interval $[a, b]$ in Example 2.1.1 by the compact set K. Hence we need only to show that (f_n^∞) is a Cauchy sequence in $(C(K), D)$.

Let $\varepsilon > 0$ be given. By the equicontinuity of the family of functions $\mathcal{K}$, we can choose $\delta > 0$ such that

$$\forall f \in \mathcal{K} \ \ \forall x, y \in X : d_X(x, y) < \delta \ \ \Rightarrow \ \ |f(x) - f(y)| \le \frac{\varepsilon}{3} \ .$$

Next choose $m > 1/\delta$. Since W_m contains only finitely many points, and all the complex sequences $(f_n^\infty(x_{mj}))$ converges, we may choose the integer n_0 so large that

$$|f_{n+p}^\infty(x_{mj}) - f_n^\infty(x_{mj})| \le \frac{\varepsilon}{3} \ \ \text{for all} \ \ x_{mj} \in W_m \ \ \text{and} \ \ n \ge n_0 \ .$$

Then for an arbitrary point $x \in K$, there exists a point $x_{mj} \in W_m$ with $d(x, x_{mj}) < 1/m < \delta$ such that whenever $n \ge n_0$

$$
\begin{aligned}
|f_{n+p}^\infty(x) - f_n^\infty(x)| &\le |f_{n+p}^\infty(x) - f_{n+p}^\infty(x_{mj})| \\
&\quad + |f_{n+p}^\infty(x_{mj}) - f_n^\infty(x_{mj})| \\
&\quad + |f_n^\infty(x_{mj}) - f_n^\infty(x)| \le \varepsilon.
\end{aligned}
$$

From this we get that $D(f_{n+p}^\infty, f_n^\infty) \le \varepsilon$ for $n \ge n_0$, proving that the sequence (f_n^∞) is a Cauchy sequence in $(C(K), D)$, and hence, as mentioned, that the subsequence (f_n^∞) of (f_n) is convergent in $(C(K), D)$.

This completes the proof that $\mathcal{K}$ is a compact subset of $(C(K), D)$.   □

1.4   The Banach Fixed Point Theorem

As an illustration of the use of completeness we prove in this section an important fixed point theorem for contracting mappings in complete metric spaces. In its final form, the theorem was formulated by Banach.

Definition 1.4.1. Let (M, d) be a metric space. A mapping $T : M \to M$ is called a *contraction* if there exists a real number λ with $0 \le \lambda < 1$, such that

$$d(Tx, Ty) \le \lambda d(x, y) \qquad \text{for all} \ \ x, y \in M.$$

The number λ is called a *contraction factor*.

Remark 1.4.2. To simplify notation one often writes Tx for the image of a point $x \in M$ under a contraction $T : M \to M$ instead of the usual $T(x)$.

Clearly a contraction $T : M \to M$ is continuous.

Together with the contraction $T : M \to M$ we shall consider its *iterates*. By this we understand the family of mappings $T^n : M \to M$ defined inductively by setting $T^2 = T \circ T$, and in general $T^n = T \circ T^{n-1}$. Note that if T is a contraction with contraction factor $\lambda \in [0, 1[$, then T^n is a contraction with contraction factor $\lambda^n \in [0, 1[$, since for all $x, y \in M$ we have the inequalities

$$\begin{aligned}
d\left(T^n x, T^n y\right) &= d\left(T(T^{n-1}x), T(T^{n-1}y)\right) \\
&\leq \lambda d(T^{n-1}x, T^{n-1}y) \\
&\vdots \\
&\leq \lambda^n d(x, y).
\end{aligned}$$

A *fixed point* for $T : M \to M$ is a point $x_0 \in M$ satisfying $Tx_0 = x_0$.

Now we are ready to prove the fixed point theorem for contractions in complete metric spaces.

Theorem 1.4.3 (The Banach Fixed Point Theorem). *In a complete metric space (M, d), any contraction $T : M \to M$ has exactly one fixed point $x_0 \in M$. If $x \in M$ is any point in M, then*

$$T^n x \to x_0 \quad \text{for} \quad n \to \infty.$$

Proof. First we prove uniqueness. Assume for that purpose that x_0 and x_1 are fixed points for T. Then we have

$$0 \leq d(x_0, x_1) = d(Tx_0, Tx_1) \leq \lambda d(x_0, x_1).$$

Since $\lambda \in [0, 1[$, this can only be satisfied when $d(x_0, x_1) = 0$, and hence $x_0 = x_1$. Therefore the contraction T has at most one fixed point.

Now to existence. Let $x \in M$ be an arbitrarily chosen point in M and consider the sequence $(T^n x)$ in M.

For any $n \in \mathbb{N}$, we have

$$d(T^n x, T^{n+1} x) = d\left(T^n x, T^n(Tx)\right) \leq \lambda^n d(x, Tx),$$

since T^n is a contraction with contraction factor $\lambda^n \in [0, 1[$.

Making repeated use of the triangle inequality followed by the above inequality, we then get, for arbitrary $n, k \in \mathbb{N}$:

$$d(T^n x, T^{n+k} x) \leq d(T^n x, T^{n+1} x) + d(T^{n+1} x, T^{n+k} x)$$

$$\vdots$$

$$\leq d(T^n x, T^{n+1} x) + d(T^{n+1} x, T^{n+2} x) + \dots$$
$$+ d(T^{n+k-1} x, T^{n+k} x)$$
$$\leq (\lambda^n + \lambda^{n+1} + \dots + \lambda^{n+k-1}) d(x, Tx)$$
$$= \lambda^n \frac{1 - \lambda^k}{1 - \lambda} d(x, Tx)$$
$$\leq \frac{\lambda^n}{1 - \lambda} d(x, Tx).$$

Since $\lambda \in [0, 1[$, clearly

$$\frac{\lambda^n}{1 - \lambda} d(x, Tx) \to 0 \quad \text{for} \quad n \to \infty.$$

Hence the above inequality for $d(T^n x, T^{n+k} x)$ shows that $(T^n x)$ is a Cauchy sequence in (M, d). Since (M, d) is a complete metric space, there exists therefore a point $x_0 \in M$ such that

$$T^n x \to x_0 \quad \text{for} \quad n \to \infty.$$

By the continuity of T it follows that

$$T^{n+1} x \to T x_0 \quad \text{for} \quad n \to \infty.$$

This shows that both x_0 and $T x_0$ are limit points for the sequence $(T^n x)$. In a metric space, a sequence can have at most one limit point and therefore $T x_0 = x_0$. Hence the point $x_0 \in M$ is indeed a fixed point for T, and thus the contraction T has at least one fixed point.

From the above it follows that the contraction T has exactly one fixed point $x_0 \in M$. Furthermore, it follows that $x_0 \in M$ can be determined as the limit point for the sequence $(T^n x)$ generated by an arbitrary point $x \in S$. This completes the proof of Theorem 1.4.3. □

1.5 The Baire Category Theorem

The following theorem is an important tool in applications of completeness in functional analysis. It was proved in 1899 by the French mathematician

René-Louis Baire (1874–1932), and is known as the *Baire Category Theorem*, or simply, as *Baire's Theorem*.

Theorem 1.5.1 (Baire's Theorem). *If (M, d) is a complete metric space, then the intersection of every countable collection of dense open subsets of M is dense in M.*

Proof. Suppose $U_1, U_2, U_3, \ldots$ is an arbitrary countable collection of dense open subsets in M.

Let $W \neq \emptyset$ be an arbitrary non-empty, open set in M. To prove that the intersection $\cap_{i=1}^{\infty} U_i$ is dense in M, we only have to show that $\cap_{i=1}^{\infty} U_i \cap W \neq \emptyset$.

Let $\{B_r(x)\}$ and $\{\overline{B_r(x)}\}$ be respectively the system of open balls and the system of the closures of the open balls in (M, d).

Since U_1 is dense in M, the intersection $U_1 \cap W$ is a non-empty, open subset of M, and hence we can choose $x_1 \in M$ and $r_1 \in \mathbb{R}^+$ such that

$$\overline{B_{r_1}(x_1)} \subseteq U_1 \cap W \quad \text{and} \quad 0 < r_1 < 1.$$

Since $B_{r_1}(x_1)$ is also a non-empty, open set and U_2 is dense in M, the subset $U_2 \cap B_{r_1}(x_1)$ is a non-empty, open subset of M and hence we can choose $x_2 \in M$ and $r_2 \in \mathbb{R}^+$ such that

$$\overline{B_{r_2}(x_2)} \subseteq U_2 \cap B_{r_1}(x_1) \quad \text{and} \quad 0 < r_2 < \frac{1}{2}.$$

Proceeding by induction we can for each $n \geq 2$ choose $x_n \in M$ and $r_n \in \mathbb{R}^+$ such that

$$\overline{B_{r_n}(x_n)} \subseteq U_n \cap B_{r_{n-1}}(x_{n-1}) \quad \text{and} \quad 0 < r_n < \frac{1}{n}.$$

The sequence (x_n) we obtain in this way is easily seen to be a Cauchy sequence in (M, d), since by construction of the sequence we have for an arbitrary $n_0 \geq 1$,

$$d(x_n, x_m) \leq d(x_n, x_{n_0}) + d(x_{n_0}, x_m) < \frac{1}{n_0} + \frac{1}{n_0} < \frac{2}{n_0},$$

for all $n, m \geq n_0$.

Since (M, d) is a complete metric space, the Cauchy sequence (x_n) has a limit point $x \in M$. The limit point x must lie in all of the closed sets $\overline{B_{r_n}(x_n)}$, and hence in all of the open sets U_n, since points x_i in the sequence lies in the closed set $\overline{B_{r_n}(x_n)}$ if $i > n$. It follows that the limit point $x \in \cap_{i=1}^{\infty} U_i \cap W$, which therefore is non-empty. This completes the proof. $\qquad\square$

1.6   Normed vector spaces

We assume that the reader has basic knowledge of the theory of vector spaces with the set of real numbers $\mathbb{R}$, or the set of complex numbers $\mathbb{C}$, as scalar field. We shall be working with vector spaces over these scalar fields as appropriate. In general we take the complex numbers as scalar field.

If the complex vector space V has dimension n, and $\{e_1, \ldots, e_n\}$ is a basis for V, then any vector $v \in V$ can be written in a unique way as a linear combination

$$v = \alpha_1 e_1 + \cdots + \alpha_n e_n, \quad \text{for} \quad \alpha_1, \ldots, \alpha_n \in \mathbb{C}.$$

The complex numbers $\alpha_1, \ldots, \alpha_n$ are called the *coordinates* of the vector in the given basis for V.

The linear mapping that associates the set of coordinates $(\alpha_1, \ldots, \alpha_n) \in \mathbb{C}^n$ to the vector $v \in V$, defines an isomorphism of V onto $\mathbb{C}^n$. From this we conclude: *Two finite dimensional complex vector spaces of the same dimension n are isomorphic, and $\mathbb{C}^n$ is a model for the isomorphism class.*

Functional analysis is mostly concerned with normed vector spaces.

A norm on a complex vector space V is first of all a real-valued function $||\cdot|| : V \to \mathbb{R}$ which to any vector $x \in V$ associates a real number $||x||$. It has proved fruitful to require that the quantity $||\cdot||$ has the following three properties enjoyed by the Euclidean norm introduced on p. 2:

NORM 1 (positive definite)
$$||x|| \geq 0, \text{ for all } x \in V, \quad = 0 \iff x = 0.$$

NORM 2 (uniform scaling)
$$||\alpha x|| = |\alpha| \, ||x|| \text{ for all } x \in V, \text{ and } \alpha \in \mathbb{C}.$$

NORM 3 (the triangle inequality)
$$||x + y|| \leq ||x|| + ||y|| \text{ for all } x, y \in V.$$

Definition 1.6.1. Let V be a complex vector space. A function $||\cdot|| : V \to \mathbb{R}$ with the properties NORM 1, NORM 2 and NORM 3 is called a *norm* in the vector space.

The pair $(V, ||\cdot||)$ consisting of a vector space V together with a specific norm $||\cdot||$ is called a *normed vector space*.

In a canonical way every norm in a complex vector space induces a metric in the vector space. This is emphasized in the following theorem.

Theorem 1.6.2. *Let V be a normed vector space with norm $||\cdot||$. Then V has a canonical structure as a metric space with the metric d given by*

$$d(x,y) = ||x-y|| \quad \text{for all } x,y \in V.$$

Proof.   According to NORM 1 we have $||x-y|| \geq 0$, and $||x-y|| = 0$ if and only if $x = y$, proving that d satisfies MET 1. Using NORM 2 we get:

$$d(x,y) = ||x-y|| = ||(-1)(y-x)||$$
$$= |-1|\,||y-x|| = ||y-x|| = d(y,x),$$

proving that d satisfies MET 2. Finally, MET 3 follows from NORM 3:

$$d(x,z) = ||x-z|| = ||(x-y)+(y-z)||$$
$$\leq ||x-y|| + ||y-z||$$
$$= d(x,y) + d(y,z).$$

This proves Theorem 1.6.2.   □

When a normed vector space is considered as a metric space, it is with the metric structure defined in Theorem 1.6.2.

In a finite dimensional normed vector space, the topology induced by the metric structure associated with the norm is independent of the norm. In other words, two norms $||\cdot||$ and $||\cdot||'$ in a finite dimensional vector space V define the same system of open sets. We state this surprising result, which is not true in infinite dimensions, without proof.

Theorem 1.6.3. *In a finite dimensional vector space all norms induce the same topology.*

In the following chapters we shall meet several normed vector spaces defined by spaces of real-valued or complex-valued functions. Here is a basic example.

Example 1.6.4. Let $[a,b]$ be a closed and bounded interval in $\mathbb{R}$, and let $C([a,b])$ be the set of continuous, complex-valued functions $f : [a,b] \to \mathbb{C}$.

For $f,g \in C([a,b])$ and $\alpha \in \mathbb{C}$ define sum and multiplication by scalar of functions by the usual pointwise defined operations

$$(f+g)(x) = f(x)+g(x) \quad \text{and} \quad (\alpha \cdot f)(x) = \alpha \cdot f(x)$$

for $x \in [a,b]$. It is easy to prove that $f+g$ and $\alpha \cdot f$ are continuous functions in $[a,b]$. Hence $C([a,b])$ gets the structure of a complex vector space (of infinite dimension) by these operations.

Since every continuous function $f : [a, b] \to \mathbb{C}$ is bounded we can define

$$\|f\| = \sup \{|f(x)| \mid x \in [a, b]\}.$$

It is easy to prove that $\|\cdot\|$ is a norm in $C([a, b])$.

The normed vector space in Example 1.6.4 (see also Example 2.1.1) is a complete metric space when equipped with the metric structure associated with the norm. Such normed vector spaces were discovered to be of central importance in functional analysis by Banach and now carry his name.

Definition 1.6.5. A normed vector space which is complete as a metric space is called a *Banach space*.

In finite dimensions there are no problems with completeness. Making appropriate use of compactness one can prove the following theorem.

Theorem 1.6.6. *Every finite dimensional normed vector space is a Banach space.*

1.7 Bounded linear operators

Applications of functional analysis often involve operations by linear mappings on elements in normed vector spaces with values in other normed vector spaces. In such contexts, a linear mapping is usually termed a linear operator, or sometimes just an *operator*.

The basic facts about continuity of linear operators between normed vector spaces are collected in the following theorem.

Theorem 1.7.1. *Let $T : V \to W$ be a linear operator between normed vector spaces $(V, \|\cdot\|_V)$ and $(W, \|\cdot\|_W)$. Then the following statements are equivalent:*

(1) *T is continuous.*

(2) *T is continuous at $0 \in V$.*

(3) *There is a constant k, such that $\|T(x)\|_W \leq k$
    for all unit vectors $x \in V$, i.e. for $\|x\|_V = 1$.*

(4) *There is a constant k, such that $\|T(x)\|_W \leq k\|x\|_V$ for all $x \in V$.*

Proof. We prove the closed cycle of implications $(1) \Rightarrow (2) \Rightarrow (3) \Rightarrow (4) \Rightarrow (1)$.
$(1) \Rightarrow (2)$. Trivial.
$(2) \Rightarrow (3)$. Suppose that T is continuous at $0 \in V$. Corresponding to $\varepsilon = 1$, choose a $\delta > 0$, such that $\|T(x)\|_W \leq 1$ for $\|x\|_V \leq \delta$.

For any unit vector $x \in V$, i.e. $||x||_V = 1$, clearly $||\delta x||_V = \delta$, and hence $||T(\delta x)||_W \leq 1$ by the choice of δ. Since T is linear we then get $\delta ||T(x)||_W \leq 1$, and consequently $||T(x)||_W \leq k$ with $k = 1/\delta$.

(3)$\Rightarrow$(4). An arbitrary vector $x \in V$ can be written in the form $x = ||x||_V x_0$ with $||x_0||_V = 1$. For a constant k satisfying (3), we then have:

$$||T(x)||_W = ||T(||x||_V x_0)||_W = ||\ ||x||_V T(x_0)||_W$$
$$= ||x||_V ||T(x_0)||_W \leq k||x||_V.$$

(4)$\Rightarrow$(1). If there exists a constant k satisfying (4), then for an arbitrary pair of vectors $x, y \in V$ we have:

$$||T(x) - T(y)||_W = ||T(x - y)||_W \leq k||x - y||_V.$$

This inequality immediately shows that T is continuous. $\square$

The equivalent properties (3) and (4) of a continuous linear operator between normed vector spaces stated in Theorem 1.7.1 inspire the definition of the important notion of a bounded linear operator.

For any normed vector space $(V, ||\cdot||_V)$ denote by

$$C(V) = \{x \in V \mid ||x||_V \leq 1\}$$

the closed *unit ball* in V. Then the definition of boundedness of an operator can be formulated as follows.

Definition 1.7.2. A linear operator $T : V \to W$ between normed vector spaces $(V, ||\cdot||_V)$ and $(W, ||\cdot||_W)$ is called a *bounded* linear operator if it maps the closed unit ball $C(V)$ in V into a bounded set (a ball) in W.

The main result recorded in Theorem 1.7.1 can then be stated as follows.

Corollary 1.7.3. *A linear operator $T : V \to W$ between normed vector spaces $(V, ||\cdot||_V)$ and $(W, ||\cdot||_W)$ is continuous if and only if it is bounded.*

Many applications of functional analysis depend on finding suitable estimates of quantities defined by bounded linear operators $T : V \to W$ between normed vector spaces in terms of a norm associated with such operators.

The norm of a bounded linear operator $T : V \to W$ measures, informally speaking, how much T distorts the unit sphere in V. The distortion of the unit sphere can be measured in several equivalent ways of which the most important are listed in the following lemma.

Lemma 1.7.4. *Let* $T : V \to W$ *be a bounded linear operator between normed vector spaces* $(V, ||\cdot||_V)$ *and* $(W, ||\cdot||_W)$. *Then each of the real numbers defined below exists:*

$$r = \inf \left\{ k \in \mathbb{R} \mid ||T(x)||_W \le k \quad \text{for all vectors } x \in V \text{ with } ||x||_V = 1 \right\}$$
$$s = \inf \left\{ k \in \mathbb{R} \mid ||T(x)||_W \le k||x||_V \quad \text{for all } x \in V \right\}$$
$$t = \sup \left\{ ||T(x)||_W \mid x \in V, \; ||x||_V = 1 \right\}$$
$$u = \sup \left\{ ||T(x)||_W \mid x \in V, \; ||x||_V \le 1 \right\},$$

and $r = s = t = u$.

Proof. Since $T : V \to W$ is bounded it follows by Theorem 1.7.1 that there exists a constant $k \in \mathbb{R}$, such that $||T(x)||_W \le k$ for all $x \in V$ with $||x||_V = 1$. Thus the set of real numbers

$$A = \left\{ k \in \mathbb{R} \mid ||T(x)||_W \le k \quad \text{for all } x \in V \text{ with } ||x||_V = 1 \right\},$$

is non-empty. Furthermore, it is bounded below by 0. Hence the number $r = \inf A$ exists. The inequality $||T(x)||_W \le k$ for all $x \in V$ with $||x||_V = 1$ immediately shows the existence of the number t.

An arbitrary vector $x \in V$ can be written in the form $x = ||x||_V x_0$ with $||x_0||_V = 1$. When x is represented in this form, we get $||T(x)||_W = ||x||_V ||T(x_0)||_W$. This shows that $||T(x_0)||_W \le k$ for all $x_0 \in V$ with $||x_0||_V = 1$ if and only if $||T(x)||_W \le k||x||_V$ for all $x \in V$. From this, it follows that

$$A = \left\{ k \in \mathbb{R} \mid ||T(x)||_W \le k||x||_V \quad \text{for all } x \in V \right\}.$$

This shows that the number s exists, and furthermore, that $r = s$. The expression $||T(x)||_W = ||x||_V ||T(x_0)||_W$ also shows immediately that, since the number t exists, the number u exists, and that $t = u$.

What is left to prove is that $r = t$. Indeed, since $||T(x)||_W \le t$ for all $x \in E$ with $||x||_V = 1$, the number t belongs to the set A over which the infimum is taken when determining r. Hence we have $r \le t$. At the same time $t \le k$ for all $k \in A$, from which it follows that $t \le r$. All in all $r = t$.

This completes the proof of Lemma 1.7.4. $\qquad\square$

Based on Lemma 1.7.4 we introduce the following definition.

Definition 1.7.5. By *the operator norm*, or simply *the norm*, of a bounded linear operator $T : V \to W$ between normed vector spaces $(V, ||\cdot||_V)$ and

$(W, ||\cdot||_W)$, we understand the real number $||T||$ determined in one of the following equivalent ways:

$$
\begin{aligned}
||T|| &= \inf\left\{k \in \mathbb{R} \mid ||T(x)||_W \leq k \quad \text{for all } x \in V \text{ with } ||x||_V = 1\right\} \\
&= \inf\left\{k \in \mathbb{R} \mid ||T(x)||_W \leq k||x||_V \quad \text{for all } x \in V\right\} \\
&= \sup\left\{||T(x)||_W \mid x \in V, \ ||x||_V = 1\right\} \\
&= \sup\left\{||T(x)||_W \mid x \in V, \ ||x||_V \leq 1\right\}.
\end{aligned}
$$

The following estimate involving the operator norm of a bounded linear operator is crucial in many situations.

Lemma 1.7.6. *Let $T : V \to W$ be a bounded linear operator between normed vector spaces $(V, ||\cdot||_V)$ and $(W, ||\cdot||_W)$. Then*

$$||T(x)||_W \leq ||T||\,||x||_V \quad \text{for all} \quad x \in V.$$

Proof. An arbitrary vector $x \in V$ can be written in the form $x = ||x||_V x_0$ with $||x_0||_V = 1$. From the definition of the operator norm it follows immediately that $||T(x_0)||_W \leq ||T||$. Then we get

$$||T(x)||_W = ||x||_V ||T(x_0)||_W \leq ||T||\,||x||_V,$$

as claimed. $\square$

For an arbitrary pair of normed vector spaces V and W, we denote by

$$B(V, W) \quad \text{the space of bounded linear operators } T : V \to W.$$

For $T, T_1, T_2 \in B(V, W)$ and $\alpha \in \mathbb{C}$ we define $T_1 + T_2$ and αT by the usual pointwise defined operations

$$
\begin{aligned}
(T_1 + T_2)(x) &= T_1(x) + T_2(x) \\
(\alpha T)(x) &= \alpha T(x)
\end{aligned}
$$

for $x \in V$.

The mappings $T_1 + T_2$ and αT are bounded linear operators and hence belong to $B(V, W)$, which thereby gets the structure of a complex vector space. But $B(V, W)$ has a richer structure. The following theorem, which is left to the reader as an exercise, justifies in particular the name operator norm.

Theorem 1.7.7. *For arbitrary bounded linear operators $T, T_1, T_2 \in B(V, W)$ and an arbitrary scalar $\alpha \in \mathbb{C}$ we have that*

NORM 1 $||T|| \geq 0$, *and* $||T|| = 0 \iff T = 0$.

NORM 2 $\|\alpha T\| = |\alpha| \|T\|$.

NORM 3 $\|T_1 + T_2\| \leq \|T_1\| + \|T_2\|$.

In other words: The operator norm is a norm in the vector space $B(V, W)$ and induces the structure of a normed vector space in this space.

If $U, V,$ and W are normed vector spaces, and $S : U \to V$ and $T : V \to W$ are bounded linear operators, it furthermore holds that

$$\|T \circ S\| \leq \|T\| \|S\|.$$

1.8 The Open Mapping Theorem

If a bounded linear operator between Banach spaces has an inverse defined on the full image space, then the inverse linear operator is automatically bounded. This is a consequence of a famous theorem proved by Banach.

For good reasons the theorem is known as the Open Mapping Theorem. First a definition.

Definition 1.8.1. A mapping $f : X \to Y$ between metric spaces, or more generally topological spaces, is said to be an *open mapping* if every open subset U in X is mapped into an open subset $f(U)$ in Y.

Theorem 1.8.2 (Open Mapping Theorem). *Let $(V, \|\cdot\|_V)$ and $(W, \|\cdot\|_W)$ be Banach spaces. Then every surjective, bounded linear operator $T : V \to W$ is an open mapping.*

Proof. Due to linearity of $T : V \to W$ and continuity of the vector space operations, the theorem amounts to proving the following

Assertion: There is an open ball $B_r^W(0)$ in W such that $B_r^W(0) \subseteq T(B_1^V(0))$.

Since T is surjective it follows that $W = \cup_{k=1}^{\infty} T(B_k^V(0)) = \cup_{k=1}^{\infty} \overline{T(B_k^V(0))}$, and hence that $\cap_{k=1}^{\infty}(W \setminus \overline{T(B_k^V(0))}) = \emptyset$.

Since $(W, \|\cdot\|_W)$ is a complete metric space, Baire's Theorem 1.5.1 shows that for some k, the open set $W \setminus \overline{T(B_k^V(0))}$ is not dense in W. This implies that the closed set $\overline{T(B_k^V(0))}$ must contain interior points, i.e. there exists a point $y_0 \in \overline{T(B_k^V(0))}$ and a radius $s > 0$ such that $B_s^W(y_0) \subseteq \overline{T(B_k^V(0))}$.

For every $z \in B_1^W(0)$, the vectors $y_0, y_0 + sz \in B_s^W(y_0) \subseteq \overline{T(B_k^V(0))}$.

By continuity of the vector space operations and linearity, the difference vector $sz \in B_s^W(0)$ satisfies

$$sz \in \overline{T(B_k^V(0))} - \overline{T(B_k^V(0))} \subseteq \overline{T(B_k^V(0)) - T(B_k^V(0))} \subseteq \overline{T(B_{2k}^V(0))} \ .$$

From this follows by linearity that

$$\forall z \in B_1^W(0) : z \in \overline{T((2k/s)||z||_W B_1^V(0))},$$

which proves

$(*) \; \forall z \in B_1^W(0) \; \forall \eta > 0 \; \exists x \in V : ||x||_V \leq (2k/s)||z||_W \;$ and $\; ||z - T(x)||_W < \eta.$

Put $r = s/2k$. Fix an arbitrary point $y \in B_r^W(0)$ and let $\varepsilon > 0$.

Using $(*)$ with $z = y$ and $\eta = 2^{-1}r\varepsilon$ we can choose an element $x_1 \in V$ for which

$$||y - T(x_1)||_W < 2^{-1}r\varepsilon \quad \text{and} \quad ||x_1||_V < 1.$$

Taking $z = y - T(x_1)$ and $\eta = 2^{-2}r\varepsilon$ in $(*)$ we can then choose an element $x_2 \in V$ such that

$$||y - T(x_1) - T(x_2)||_W < 2^{-2}r\varepsilon \quad \text{and} \quad ||x_2||_V < 2^{-1}\varepsilon.$$

In this manner, using $(*)$ we can proceed by induction and choose elements $x_n \in V$ for all $n \geq 2$ such that

$$||y - T(x_1) - \cdots - T(x_n)||_W < 2^{-n}r\varepsilon \quad \text{and} \quad ||x_n||_V < 2^{-(n-1)}\varepsilon.$$

The sequence $(\sigma_n = x_1 + \cdots + x_n)$ is a Cauchy sequence in $(V, ||\cdot||_V)$, which follows by noticing that

$$||\sigma_m - \sigma_n||_V \leq \sum_{i=n+1}^{\infty} 2^{-i}\varepsilon = 2^{-n}\varepsilon \quad \text{for} \quad 1 \leq n < m.$$

Since $(V, ||\cdot||_V)$ is complete, there exists an $x \in V$ such that $\sigma_n \to x$ for $n \to \infty$. The element x has norm $||x||_V < 1 + \varepsilon$, since $\sum_{i=1}^{\infty} 2^{-i} = 1$

Since T is continuous, we get $T(\sigma_n) \to T(x)$ for $n \to \infty$, and by construction of the sequence (σ_n), we have $T(\sigma_n) \to y$ for $n \to \infty$. We conclude that $T(x) = y$.

Since $y \in B_r^W(0)$ was arbitrarily chosen, we have now proved that

$$rB_1^W(0) \subseteq T((1+\varepsilon)B_1^V(0))$$

or

$$(1+\varepsilon)^{-1}rB_1^W(0) \subseteq T(B_1^V(0)).$$

By letting $\varepsilon \to 0$ and taking the union of the balls on the left-hand side of the inclusion, we get in the limit that $B_r^W(0) \subseteq T(B_1^V(0))$. This completes the proof of the assertion and hence the proof of the theorem. $\quad\square$

The Open Mapping Theorem has the following important corollary which is generally known as Banach's Theorem.

Theorem 1.8.3 (Banach's Theorem). *Let $(V, ||\cdot||_V)$ and $(W, ||\cdot||_W)$ be Banach spaces. Then the inverse linear mapping $T^{-1} : W \to V$ to a bijective, bounded linear mapping $T : V \to W$ is also bounded.*

Proof. Let U be an arbitrary open set in $(V, ||\cdot||_V)$. Then $(T^{-1})^{-1}(U) = T(U)$ is an open set in $(W, ||\cdot||_W)$ by the Open Mapping Theorem 1.8.2. By Theorem 1.2.3 this shows that T^{-1} is continuous and hence a bounded linear mapping by Corollary 1.7.3.   □

Another important corollary of the Open Mapping Theorem is known as the Closed Graph Theorem. We prepare the statement of the theorem by some definitions.

If $(V, ||\cdot||_V)$ and $(W, ||\cdot||_W)$ is a pair of normed vector spaces, we define the *product* $V \otimes W$ of normed vector spaces as the product set

$$V \otimes W = \{(x, y) \, | \, x \in V, y \in W\} \, ,$$

equipped with the coordinatewise defined vector space structure and the norm

$$||(x, y)||_1 = ||x||_V + ||y||_W \quad \text{for} \quad x \in V, \ y \in W \, .$$

It is easy to prove that $(V \otimes W, ||\cdot||_1)$ is a Banach space if $(V, ||\cdot||_V)$ and $(W, ||\cdot||_W)$ are Banach spaces.

If $T : V \to W$ is a linear operator, the *graph* $G(T)$ of T is the linear subspace in $(V \otimes W, ||\cdot||_1)$ defined by

$$G(T) = \{(x, T(x)) \, | \, x \in V\} \, .$$

Theorem 1.8.4 (Closed Graph Theorem). *A linear mapping $T : V \to W$ between Banach spaces $(V, ||\cdot||_V)$ and $(W, ||\cdot||_W)$ is bounded if and only if the graph $G(T)$ of T is a closed linear subspace in $(V \otimes W, ||\cdot||_1)$.*

Proof. First suppose that $T : V \to W$ is a bounded linear mapping. Since T is bounded, a convergent sequence (x_n) in V with limit point $x \in V$, is mapped into a convergent sequence $(T(x_n))$ in W with limit point $T(x) \in W$. Therefore, if $(x_n, T(x_n))$ is a sequence in $G(T)$ that converges to a point $(x, y) \in V \otimes W$, it follows that (x_n) converges to x and hence that $(T(x_n))$ converges to $T(x)$, proving that $(x, y) = (x, T(x)) \in G(T)$. This shows that $G(T)$ contains all of its contact points, and hence that it is a closed linear subspace in $(V \otimes W, ||\cdot||_1)$.

Next suppose that $G(T)$ is a closed linear subspace of $(V \otimes W, ||\cdot||_1)$. Then $G(T)$ is a Banach space, since $V \otimes W$ is a Banach space. Let $P_1 : G(T) \to V$

be the projection of $G(T)$ onto V, i.e. $P_1(x, T(x)) = x$ for all $x \in V$, and $P_2 : G(T) \to W$ the projection of $G(T)$ into W, i.e. $P_2(x, T(x)) = T(x)$ for all $x \in V$. Clearly, P_1 is a bijective, bounded linear mapping between Banach spaces, and hence the inverse linear mapping $P_1^{-1} : V \to G(T)$ is bounded according to Banach's Theorem 1.8.3. Since P_2 is also a bounded linear mapping, and since $T = P_2 P_1^{-1}$, it follows that $T : V \to W$ is a bounded linear mapping. $\qquad\square$

1.9   The Hahn-Banach Theorem

In this section we shall state and prove the Hahn-Banach Theorem, which together with the Open Mapping Theorem and the Closed Graph Theorem are the three fundamental theorems in functional analysis of linear operators. The proof of the Hahn-Banach Theorem is more delicate than the proofs of the other two theorems, since it needs the Axiom of Choice. Mathematicians most often accept the Axiom of Choice straightaway, and thereby they also accept the equivalent statement known as

Zorn's Lemma If $\mathcal{P}$ is an ordered set with the partial ordering $\prec$, and if every totally ordered subset $\mathcal{T}$ of $\mathcal{P}$ has an upper bound in $\mathcal{P}$, then $\mathcal{P}$ has a maximal element.

(A subset $\mathcal{T}$ of $\mathcal{P}$ is called *totally ordered*, if each two elements $A, B \in \mathcal{T}$ satisfy either $A \prec B$ or $B \prec A$.)

The proof of the Hahn-Banach Theorem requires separate treatments of the real and the complex case. The general theorem was proved with an appeal to Zorn's Lemma by the Austrian mathematician Hans Hahn (1879–1934) in 1927 and Stefan Banach in 1929.

Theorem 1.9.1 (Hahn-Banach Theorem). *Let $(V, ||\cdot||_V)$ be a normed vector space over $\mathbb{K} = \mathbb{R}, \mathbb{C}$, and let U be an arbitrary subspace of V.*

Then every bounded linear functional $f : U \to \mathbb{K}$ admits an extension to a bounded linear functional $F : V \to \mathbb{K}$ so that $||F|| = ||f||$.

Proof.   A bounded linear functional $F : V \to \mathbb{K}$ is called an extension of the bounded linear functional $f : U \to \mathbb{K}$, if $F(x) = f(x)$ for all $x \in U$. The operator norms of the functionals are given by

$$||F|| = \sup_{x \in V} \{ |F(x)| \mid ||x||_V = 1 \}$$
$$||f|| = \sup_{x \in U} \{ |f(x)| \mid ||x||_V = 1 \}.$$

First we consider the case $\mathbb{K} = \mathbb{R}$, i.e. the vector spaces are real vector spaces, and the bounded linear functionals are real-valued.

If $U = V$, there is nothing to prove. Hence suppose $V \setminus U \neq \emptyset$, and let $x_0 \in V \setminus U$ be an arbitrary element.

Let V_1 be the subspace of V spanned by U and x_0, i.e.

$$V_1 = \{x = u + tx_0 \,|\, u \in U,\ t \in \mathbb{R}\} \ .$$

It is easy to prove that the expansion $x = u + tx_0$ of an element $x \in V_1$ is unique. For any real number $c \in \mathbb{R}$, we can therefore define a function $f_1 : V_1 \to \mathbb{R}$ by

$$f_1(u + tx_0) = f(u) + ct, \quad u \in U,\ t \in \mathbb{R} \ .$$

Clearly $f_1 : V_1 \to \mathbb{R}$ is a linear functional for which $f_1(x) = f(x)$ for all $x \in U$.

We shall now prove that we can choose $c \in \mathbb{R}$ such that f_1 is a bounded linear functional with $\|f_1\| \leq \|f\|$. This will imply that $\|f_1\| = \|f\|$, since obviously $\|f\| \leq \|f_1\|$.

For arbitrary elements $u, v \in U$, we have the following estimates

$$f(u) - f(v) = f(u - v) \leq \|f\|\|u - v\| \leq \|f\|\left(\|u + x_0\| + \|x_0 + v\|\right) \ ,$$

from which we get

$$-f(v) - \|f\|\|x_0 + v\| \leq -f(u) + \|f\|\|u + x_0\| \ .$$

Since supremum of the left-hand side over $v \in U$ is less than infimum of the right-hand side over $u \in U$, we can choose $c \in \mathbb{R}$ such that

$$\sup_{v \in U}\left(-f(v) - \|f\|\|x_0 + v\|\right) \leq c \leq \inf_{u \in U}\left(-f(u) + \|f\|\|u + x_0\|\right) \ .$$

For $t > 0$, put $v = tu$. Multiplying by t on both sides of the right inequality, then adding $f(v)$ on both sides, and finally replacing v by u, we get

$$f_1(u + tx_0) = f(u) + ct \leq \|f\|\|u + tx_0\| \ .$$

For $t < 0$, put $u = tv$. Multiplying by t on both sides of the left inequality, then adding $f(u)$ on both sides, we get again

$$f_1(u + tx_0) = f(u) + ct \leq \|f\|\|u + tx_0\| \ .$$

Therefore $f_1(x) \leq \|f\|\|x\|$ for all $x = u + tx_0 \in V_1$. Replacing x by $-x$, we also get $-f_1(x) \leq \|f\|\|x\|$, which implies that $|f_1(x)| \leq \|f\|\|x\|$ for all $x \in V_1$. This proves that $\|f_1\| \leq \|f\|$, and hence as remarked, that $\|f_1\| = \|f\|$.

Now consider the set $\mathcal{P}$ of all extensions of (U, f), i.e. all pairs $(\mathcal{V}, f_{\mathcal{V}})$ consisting of a linear subspace $\mathcal{V}$ of V with $U \subseteq \mathcal{V} \subseteq V$, and a bounded linear functional $f_\mathcal{V} : \mathcal{V} \to \mathbb{R}$, such that $f_\mathcal{V}$ is an extension of f with $\|f_\mathcal{V}\| = \|f\|$. Introduce a partial ordering $\prec$ in $\mathcal{P}$ by $(\mathcal{V}', f_{\mathcal{V}'}) \prec (\mathcal{V}'', f_{\mathcal{V}''})$ if $\mathcal{V}' \subseteq \mathcal{V}''$ and $f_{\mathcal{V}''}$ is an extension of $f_{\mathcal{V}'}$.

If $\mathcal{T}$ is an arbitrary totally ordered subset of $\mathcal{P}$, i.e. each two extensions in $\mathcal{T}$ satisfy either $(\mathcal{V}', f_{\mathcal{V}'}) \prec (\mathcal{V}'', f_{\mathcal{V}''})$ or $(\mathcal{V}'', f_{\mathcal{V}''}) \prec (\mathcal{V}', f_{\mathcal{V}'})$, then $\mathcal{T}$ has an upper bound in $\mathcal{P}$, namely the pair $(\mathcal{V}_\mathcal{T}, f_{\mathcal{V}_\mathcal{T}})$, where $\mathcal{V}_\mathcal{T}$ is the union of the vector spaces in $\mathcal{T}$ and $f_{\mathcal{V}_\mathcal{T}}$ is the unique linear functional which restricts to the linear functional in every extension of (U, f) belonging to $\mathcal{T}$. Hence by Zorn's Lemma, there exists a maximal extension $(\mathcal{V}^{max}, f_{\mathcal{V}^{max}})$ for the set $\mathcal{P}$ of extensions of (U, f). We assert that $\mathcal{V}^{max} = V$. For suppose this was not the case. Then we could pick an element $x_0 \in V \setminus \mathcal{V}^{max}$ and argue as before with $\mathcal{V}^{max}$ replacing U and construct an extension (V_1, f_1) of $(\mathcal{V}^{max}, f_{\mathcal{V}^{max}})$, contradicting that $(\mathcal{V}^{max}, f_{\mathcal{V}^{max}})$ was maximal in $\mathcal{P}$. Therefore $F = f_{\mathcal{V}^{max}}$ is the desired extension of $f : U \to \mathbb{R}$ to $F : V \to \mathbb{R}$. This completes the proof of the Hahn-Banach Theorem in the real case.

Next we turn to the complex case $\mathbb{K} = \mathbb{C}$. Note here, that if we restrict multiplication with scalars to real numbers, then any complex vector space can also be considered as a real vector space.

Let $f : U \to \mathbb{C}$ be a bounded complex linear functional. If i denotes the imaginary unit, we can write $f : U \to \mathbb{C}$ as $f(x) = g(x) + ih(x)$ for $x \in U$, where $g, h : U \to \mathbb{R}$ is the real and the imaginary part of f, respectively. Clearly, $g, h : U \to \mathbb{R}$ are then bounded, real linear functionals with $|g(x)| \le |f(x)|$ and $|h(x)| \le |f(x)|$ for all $x \in U$, which implies that $\|g\| \le \|f\|$ and $\|h\| \le \|f\|$.

For each $x \in U$, we have

$$g(ix) + ih(ix) = f(ix) - if(x) = i(g(x) + ih(x)) = -h(x) + ig(x) ,$$

which implies that $h(x) = -g(ix)$ for all $x \in U$. Consequently,

$$f(x) = g(x) - ig(ix) \quad \text{for all} \quad x \in U .$$

From the real case we know that there exists a bounded, real linear functional $G : V \to \mathbb{R}$, which is an extension of $g : U \to \mathbb{R}$ and for which $\|G\| = \|g\| \le \|f\|$.

Now define $F : V \to \mathbb{C}$ by

$$F(x) = G(x) - iG(ix) \quad \text{for all} \quad x \in V .$$

It is easy to prove that $F : V \to \mathbb{C}$ is a bounded, complex linear functional. Complex linearity follows by noticing that $F(ix) = iF(x)$ for all $x \in V$, as shown by the following computation

$$F(ix) = G(ix) - iG(-x) = G(ix) + iG(x) = iF(x) .$$

For all $x \in U$, we have

$$F(x) = G(x) - iG(ix) = g(x) - ig(ix) = f(x) ,$$

showing that F is an extension of f.

For every $x \in V$, we can choose a complex root of unity $e^{i\theta} \in \mathbb{C}$ such that $e^{i\theta}F(x) = F(e^{i\theta}x)$ is a real number. Hence $F(e^{i\theta}x) = G(e^{i\theta}x)$, and we get

$$|F(x)| = F(e^{i\theta}x) = G(e^{i\theta}x) \le ||G|| \, ||e^{i\theta}x|| = ||g|| \, ||x|| \le ||f|| \, ||x|| ,$$

for all $x \in V$, proving that F is a bounded, complex linear functional with $||F|| \le ||f||$. Since trivially $||f|| \le ||F||$, we get $||F|| = ||f||$, completing the proof of the Hahn-Banach Theorem in the complex case. $\qquad \square$

The Hahn-Banach Theorem has the following immediate corollary. Again let $\mathbb{K} = \mathbb{R}, \mathbb{C}$.

Corollary 1.9.2. *For each element $x_0 \ne 0$ in a normed vector space $(V, ||\cdot||_V)$ over $\mathbb{K}$, there exists a bounded linear functional $F : V \to \mathbb{K}$ such that*

$$F(x_0) = ||x_0|| \quad and \quad ||F|| = 1 .$$

Proof. Consider the 1-dimensional linear subspace $U \subseteq V$ spanned by x_0, and let $f : U \to \mathbb{K}$ be the unique linear functional such that $f(x_0) = ||x_0||$. Then $||f|| = 1$. By the Hahn-Banach Theorem, there exists a bounded linear functional $F : V \to \mathbb{K}$ such that $F(x_0) = f(x_0) = ||x_0||$ and $||F|| = ||f|| = 1$. This completes the proof. $\qquad \square$

Chapter 2

New Types of Function Spaces

2.1 Completion of metric spaces and normed vector spaces

A *Banach space* is a normed vector space in which every Cauchy sequence is convergent. We like our spaces to have this fundamental property.

Example 2.1.1. Let $V = C([a, b])$ denote the space of continuous functions $f : [a, b] \to \mathbb{R}$ (or, $f : [a, b] \to \mathbb{C}$) defined in a closed interval $[a, b]$.
   Then V is a vector space and

$$\|f\|_\infty = \sup_{x \in [a,b]} |f(x)| \, ,$$

is a norm in V, the so-called *uniform norm*, or *supremum norm* in V.

Proposition 2.1.2. $(V, \|\cdot\|_\infty)$ *is a Banach space.*

Proof.   Let (f_n) be a Cauchy sequence in $(V, \|\cdot\|_\infty)$. Then

$$\forall \varepsilon > 0 \; \exists n_0 \in \mathbb{N} \; \forall n, m \in \mathbb{N} : \quad n, m \geq n_0 \; \Rightarrow \; \|f_n - f_m\|_\infty \leq \varepsilon.$$

This is equivalent to

$$\forall \varepsilon > 0 \; \exists n_0 \in \mathbb{N} \; \forall n, m \in \mathbb{N} : \quad n, m \geq n_0 \; \Rightarrow \; \forall x \in [a, b] : |f_n(x) - f_m(x)| \leq \varepsilon.$$

The second formulation shows that $(f_n(x))$ is a Cauchy sequence in $\mathbb{R}$ (or $\mathbb{C}$) for all $x \in [a, b]$.
   Since $(\mathbb{R}, |\cdot|)$ (or $(\mathbb{C}, |\cdot|)$) is a complete metric space it follows that

$$\forall x \in [a, b] \; \exists y \in \mathbb{R} : \quad f_n(x) \to y \; \text{ for } \; n \to \infty \, .$$

The limit y is unique for every x, and hence we can define the function

$$f : [a, b] \to \mathbb{R} \qquad \text{by} \qquad f(x) = \lim_{n \to \infty} f_n(x) \; \text{for} \; x \in [a, b] \, .$$

If we let $m \to \infty$ in the second formulation of the Cauchy condition we get

$$\forall \varepsilon > 0 \; \exists n_0 \in \mathbb{N} \; \forall n \in \mathbb{N}: \quad n \geq n_0 \; \Rightarrow \; \forall x \in [a,b] : |f_n(x) - f(x)| \leq \varepsilon.$$

Assertion 2.1.3. *The function $f : [a,b] \to \mathbb{R}$ is continuous.*

Proof. Observe that

$$|f(x) - f(x_0)| \leq |f(x) - f_n(x)| + |f_n(x) - f_n(x_0)| + |f_n(x_0) - f(x_0)| .$$

Given $\varepsilon > 0$. Choose $n_0 \in \mathbb{N}$:

$$n \geq n_0 \; \Rightarrow \; \forall x \in [a,b] : |f_n(x) - f(x)| \leq \frac{\varepsilon}{3} .$$

Then:

$$n \geq n_0 \; \Rightarrow \; |f(x) - f(x_0)| \leq \frac{\varepsilon}{3} + |f_n(x) - f_n(x_0)| + \frac{\varepsilon}{3} .$$

Now, since $f_{n_0} : [a,b] \to \mathbb{R}$ is continuous, we can (and do) choose $\delta > 0$ such that

$$|x - x_0| \leq \delta \; \Rightarrow \; |f_{n_0}(x) - f_{n_0}(x_0)| \leq \frac{\varepsilon}{3} .$$

Altogether we can conclude that

$$|x - x_0| \leq \delta \; \Rightarrow \; |f(x) - f(x_0)| \leq \varepsilon .$$

This proves that $f : [a,b] \to \mathbb{R}$ is continuous as asserted. $\qquad\square$

Now that we know $f : [a,b] \to \mathbb{R}$ is continuous, we can use the earlier proved statement

$$\forall \varepsilon > 0 \; \exists n_0 \in \mathbb{N} \; \forall n \in \mathbb{N}: \quad n \geq n_0 \; \Rightarrow \; \forall x \in [a,b] : |f_n(x) - f(x)| \leq \varepsilon ,$$

to conclude that

$$\forall \varepsilon > 0 \; \exists n_0 \in \mathbb{N} \; \forall n \in \mathbb{N}: \quad n \geq n_0 \; \Rightarrow \; \|f_n - f\|_\infty \leq \varepsilon.$$

In other words:

$$f_n \to f \quad \text{for} \quad n \to \infty \quad \text{in} \quad (V, \|\cdot\|_\infty) .$$

This completes the proof that $(V, \|\cdot\|_\infty)$ is a Banach space. $\qquad\square$

Example 2.1.4. Let $V = C([-1,1])$ denote the space of continuous functions $f : [-1,1] \to \mathbb{R}$.

Then V is a vector space and

$$\|f\|_1 = \int_{-1}^{1} |f(x)| dx \ ,$$

is a norm in V, the so-called 1-*norm* in V.

Define a sequence (f_n) of functions $f_n : [-1,1] \to \mathbb{R}$ by

$$f_n(x) = \begin{cases} 1 & \text{for} \quad -1 \le x \le 0 \\ 1 - nx & \text{for} \quad 0 \le x \le 1/n \\ 0 & \text{for} \quad 1/n \le x \le 1 \ . \end{cases}$$

By considering areas of suitable triangles defined by the graphs of the functions, we get

$$\begin{aligned} \|f_n - f_{n+k}\|_1 &= \int_{-1}^{1} |f_n(x) - f_{n+k}(x)| dx \\ &\le \int_0^{\frac{1}{n}} f_n(x) dx \\ &= \frac{1}{2} \cdot \frac{1}{n} \ \to \ 0 \quad \text{for} \quad n \to \infty \ . \end{aligned}$$

This shows that (f_n) is a Cauchy sequence in $(V, \|\cdot\|_1)$.

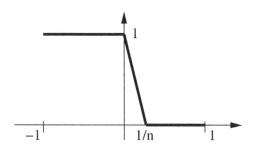

Fig. 2.1   Graph of f_n

Assertion 2.1.5. *The sequence of functions (f_n) is not convergent in $(V, \|\cdot\|_1)$ to any continuous function.*

Proof. The proof is by contradiction. Suppose that (f_n) is convergent to a continuous function $f_0 : [-1,1] \to \mathbb{R}$ in $(V, \|\cdot\|_1)$. Then by necessity the function f_0 must satisfy

$$f_0(x) = \begin{cases} 1 & \text{for } -1 \le x < 0 \\ 0 & \text{for } 0 < x \le 1 \end{cases}.$$

This is easy to prove by consideration of areas under graphs of continuous functions, since otherwise for large n, the 1-norm $\|f_n - f_0\|_1 \ge \varepsilon_0 > 0$ for a suitably small $\varepsilon_0 > 0$. Such a function cannot be continuous at $x = 0$ and we have obtained a contradiction. □

Since (f_n) is a Cauchy sequence without a limit in $(V, \|\cdot\|_1)$ we conclude that $(V, \|\cdot\|_1)$ is *not* a Banach space.

To remedy the situation in Example 2.1.4, we have to add functions to the vector space V, in this case the so-called *Lebesgue integrable* functions. In particular, we have to add functions exhibiting discontinuities such as the limit function for the sequence (f_n) in the example.

If we add these functions to the vector space V in Example 2.1.4, we obtain the space of so-called L^1-functions, $L^1([-1, 1])$. The 1-norm can be extended to this larger space such that we get a Banach space $(L^1([-1, 1]), \|\cdot\|_1)$.

There is a general principle by which we can complete metric spaces and normed vector spaces such that Cauchy sequences become convergent. The idea is simply to build a new space out of the set of Cauchy sequences, in other words, to let the Cauchy sequences represent new 'points', which will then serve as the limit points of the Cauchy sequences in the larger space.

Theorem 2.1.6 (The Completion Theorem).

(1) *Let (M, d) be a metric space. Then there exists a complete metric space (M', d') and a dense subset $\tilde{M} \subseteq M'$ ($\overline{\tilde{M}} = M'$) such that (M, d) and $(\tilde{M}, d')$ are isometric, i.e. there exists a bijective mapping $T : M \to \tilde{M}$ such that $d'(Tx, Ty) = d(x, y)$ for all $x, y \in M$.*

All completions of the metric space (M, d) are isometric.

(2) *Let $(V, \|\cdot\|)$ be a normed vector space. Then there exists a complete normed vector space (Banach space) $(V', \|\cdot\|')$ and a dense subspace $\tilde{V} \subseteq V'$ ($\overline{\tilde{V}} = V'$) such that $(V, \|\cdot\|)$ and $(\tilde{V}, \|\cdot\|')$ are isometric, i.e. there exists a bijective linear mapping (an isomorphism) $T : V \to \tilde{V}$ such that $\|T(x)\|' = \|x\|$ for all $x \in V$.*

All completions of the normed vector space $(V, \|\cdot\|)$ are isometric.

Proof.   We do the proof of (2), which is the result that will be used in the following. The proof of (1) is similar and is left to the reader.

Define V' to be the set of equivalence classes of Cauchy sequences (x_n) in $(V, ||\cdot||)$, where we identify two Cauchy sequences (x_n) and (y_n) according to the equivalence relation $\sim$:

$$(x_n) \sim (y_n) \iff ||x_n - y_n|| \to 0 \quad \text{for} \quad n \to \infty .$$

Let $[(x_n)]$ denote the equivalence class of the Cauchy sequence (x_n).

Define a vector space structure and a norm $||\cdot||'$ in V' by the following definitions

$$[(x_n)] + [(y_n)] = [(x_n + y_n)]$$
$$\alpha[(x_n)] = [(\alpha x_n)]$$
$$||[(x_n)]||' = \lim_{n\to\infty} ||x_n|| .$$

Some remarks on these definitions are appropriate.

The inequality

$$||(x_n + y_n) - (x_m + y_m)|| \le ||x_n - x_m|| + ||y_n - y_m||$$

shows that $(x_n + y_n)$ is a Cauchy sequence in $(V, ||\cdot||)$.

The inequality

$$\big| ||x_n|| - ||x_m|| \big| \le ||x_n - x_m||$$

shows that $(||x_n||)$ is a Cauchy sequence in $(\mathbb{R}, |\cdot|)$, which is a complete space. Hence $\lim_{n\to\infty} ||x_n||$ does exist.

Define the mapping

$$T : V \to V' \qquad \text{by} \qquad T(x) = [(x_n = x)] \quad \text{for all } x \in V .$$

Note that a sequence (x_n) in which an element $x \in V$ is just repeated, trivially is a Cauchy sequence in $(V, ||\cdot||)$.

It is easy to prove that the mapping T is linear, injective and that it satisfies

$$||T(x)||' = \lim_{n\to\infty} ||x|| = ||x|| .$$

Finally, we put

$$\tilde{V} = T(V) .$$

Then $\tilde{V}$ is a linear subspace of V' isomorphic (by the mapping T) to V.

The constructions made also ensure that $\tilde{V}$ is dense in V'. In fact, the element $[(x_n)] \in V'$ is the limit in $(V', ||\cdot||')$ of the sequence $(T(x_n)) \in \tilde{V}$; it boils down to the double limit $\lim_{m \to \infty}(\lim_{n \to \infty} ||\, x_n - x_m\,||) = 0$, which is an easy consequence of the Cauchy property for the sequence (x_n).

This completes the constructions involved in (2). It only remains to prove that $(V', ||\cdot||')$ is a *complete* normed vector space.

For that purpose, let (v_n) be an arbitrary Cauchy sequence in $(V', ||\cdot||')$. Since $T(V) = \tilde{V}$ is dense in V', we can choose a sequence (x_n) in V such that $||\, v_n - T(x_n)\,||' \leq 1/n$ for every $n \in \mathbb{N}$. Then $(T(x_n))$ is also a Cauchy sequence in $(V', ||\cdot||')$. Since T preserves norms, it follows that (x_n) is a Cauchy sequence in $(V, ||\cdot||)$, and hence that it defines an element $v = [(x_n)] \in V'$. Since $T(x_n) \to [(x_n)]$ for $n \to \infty$, it follows immediately that $v_n \to v$ for $n \to \infty$. This proves that $(V', ||\cdot||')$ is a complete normed vector space.

Finally, assume that $(V', ||\cdot||')$ and $(V'', ||\cdot||'')$ are two completions of the normed vector space $(V, ||\cdot||)$. Then there is an isometric, linear isomorphism from a dense subspace $\tilde{V}$ in $(V', ||\cdot||')$ onto a dense subspace $\tilde{\tilde{V}}$ in $(V'', ||\cdot||'')$. This isometric, linear isomorphism extends by continuity to an isometric, linear isomorphism of $(V', ||\cdot||')$ onto $(V'', ||\cdot||'')$.

This completes the proof of (2). $\qquad\qquad\qquad\qquad\qquad\qquad\qquad\qquad\square$

Remark 2.1.7. If the rational numbers $\mathbb{Q}$ are completed following the procedure in Theorem 2.1.6, we get the set of real numbers $\mathbb{R}$. In other words: One can think of the real numbers as equivalence classes of Cauchy sequences of rational numbers.

The elements of V' in the completion $(V', ||\cdot||')$ of $(V, ||\cdot||)$ bear a corresponding relation to the elements of V as the elements of $\mathbb{R}$ (the real numbers) bear to those of $\mathbb{Q}$ (the rational numbers). After all, the real numbers are also abstract mathematical constructs when we get down to the mathematical details about their nature.

2.2   The Weierstrass Approximation Theorem

The Completion Theorem sends a clear message that dense subspaces are important. The following theorem of Weierstrass is a spectacular result in this context.

Theorem 2.2.1 (The Weierstrass Approximation Theorem). *Let P denote the set of polynomials on a closed and bounded interval $[a, b]$. Then P is dense in $(C([a, b]), ||\cdot||_\infty)$.*

For the proof of this theorem we need a formula known from probability theory; it occurs in connection with computing the mean value and the variance of the binomial distribution.

Lemma 2.2.2. *Let n be an arbitrary positive integer and let t be a real parameter. Then*

$$\sum_{k=0}^{n}(t-\frac{k}{n})^2\binom{n}{k}t^k(1-t)^{n-k}=\frac{1}{n}t(1-t) \ .$$

Proof.   Define the function $G(s)$ in the real variable s, using the binomial formula for the rewriting of the function:

$$G(s) = \bigl(st+(1-t)\bigr)^n$$
$$= \sum_{k=0}^{n}\binom{n}{k}t^k(1-t)^{n-k}s^k \ .$$

By differentiation with respect to the parameter s, we get the formulas:

$$G'(s) = nt\bigl(st+(1-t)\bigr)^{n-1}$$
$$= \sum_{k=1}^{n}k\binom{n}{k}t^k(1-t)^{n-k}s^{k-1} \ .$$
$$G''(s) = n(n-1)t^2\bigl(st+(1-t)\bigr)^{n-2}$$
$$= \sum_{k=2}^{n}k(k-1)\binom{n}{k}t^k(1-t)^{n-k}s^{k-2} \ .$$

Then we get the function values

$$G(1) = 1 = \sum_{k=0}^{n}\binom{n}{k}t^k(1-t)^{n-k} \ .$$

$$G'(1) = nt = \sum_{k=0}^{n}k\binom{n}{k}t^k(1-t)^{n-k} \ .$$

$$G''(1) = n(n-1)t^2 = \sum_{k=0}^{n}k(k-1)\binom{n}{k}t^k(1-t)^{n-k} \ .$$

The formula in the Lemma now follows by the computations:

$$\text{Variance}/n^2 = \sum_{k=0}^{n} (t - \frac{k}{n})^2 \binom{n}{k} t^k (1-t)^{n-k}$$

$$= t^2 G(1) - \frac{2}{n} t G'(1) + \sum_{k=0}^{n} \frac{k^2}{n^2} \binom{n}{k} t^k (1-t)^{n-k}$$

$$= t^2 G(1) - \frac{2}{n} t G'(1) + \frac{1}{n^2} G''(1) + \frac{1}{n^2} G'(1)$$

$$= t^2 - \frac{2}{n} t n t + \frac{1}{n^2} n(n-1) t^2 + \frac{1}{n^2} n t$$

$$= t^2 - 2t^2 + t^2 - \frac{1}{n} t^2 + \frac{1}{n} t = \frac{1}{n} t(1-t) \ .$$

This completes the proof of Lemma 2.2.2.   □

Now we are ready for the proof of Theorem 2.2.1.

Proof.   (The Weierstrass Approximation Theorem) After a possible rescaling of the interval, it suffices to consider $[a, b] = [0, 1]$.

For an arbitrary continuous function $f : [0, 1] \to \mathbb{R}$, we define the so-called *Bernstein polynomial* $p_n(f) : [0, 1] \to \mathbb{R}$ for f:

$$p_n(f)(t) = \sum_{k=0}^{n} f(\frac{k}{n}) \binom{n}{k} t^k (1-t)^{n-k} \ .$$

Observe that

$$1 = \left(t + (1-t) \right)^n = \sum_{k=0}^{n} \binom{n}{k} t^k (1-t)^{n-k} \ .$$

The quantities $\binom{n}{k} t^k (1-t)^{n-k}$ are the densities in the binomial distribution for the number n, and in some sense one can consider $p_n(f)(t)$ as the 'weight' of f in the binomial distribution.

Since the function $f : [0, 1] \to \mathbb{R}$ is continuous in a closed and bounded interval $[0, 1]$, it is uniformly continuous. Hence

$$\forall \varepsilon > 0 \ \exists \delta > 0 \ \forall t', t'' \in [0, 1] : \ |t' - t''| \le \delta \ \Rightarrow \ |f(t') - f(t'')| \le \varepsilon \ .$$

The function $f : [0, 1] \to \mathbb{R}$ is also bounded and we put

$$M = \sup_{t \in [0,1]} |f(t)| \ .$$

Given $\varepsilon > 0$. Choose $\delta > 0$ according to uniform continuity of f. Then we have the estimates, using the formula in Lemma 2.2.2 for the final summation:

$$|f(t) - p_n(f)(t)| = \left| \sum_{k=0}^{n} \left(f(t) - f(\tfrac{k}{n}) \right) \binom{n}{k} t^k (1-t)^{n-k} \right|$$

$$\leq \sum_{|t - \frac{k}{n}| < \delta} |f(t) - f(\tfrac{k}{n})| \binom{n}{k} t^k (1-t)^{n-k}$$

$$+ \sum_{|t - \frac{k}{n}| \geq \delta} |f(t) - f(\tfrac{k}{n})| \binom{n}{k} t^k (1-t)^{n-k}$$

$$\leq \sum_{k=0}^{n} \varepsilon \binom{n}{k} t^k (1-t)^{n-k}$$

$$+ \sum_{(t - \frac{k}{n})^2 \geq \delta^2} 2M \binom{n}{k} t^k (1-t)^{n-k}$$

$$\leq \varepsilon + \frac{2M}{\delta^2} \sum_{k=0}^{n} (t - \tfrac{k}{n})^2 \binom{n}{k} t^k (1-t)^{n-k}$$

$$= \varepsilon + \frac{2M}{\delta^2} \cdot \frac{1}{n} \cdot t(1-t) \leq \varepsilon + \frac{2M}{\delta^2} \cdot \frac{1}{n} \cdot \frac{1}{4} .$$

Altogether we conclude that

$$|f(t) - p_n(f)(t)| \leq \varepsilon + \frac{2M}{\delta^2} \cdot \frac{1}{n} \cdot \frac{1}{4} , \quad \text{for all } t \in [0,1] ,$$

which implies that

$$\|f - p_n(f)\|_\infty \leq 2\varepsilon \quad \text{for} \quad n > \frac{M}{2\delta^2 \varepsilon} ,$$

and consequently that

$$p_n(f) \to f \quad \text{for} \quad n \to \infty \quad \text{in} \quad (C([0,1]), \|\cdot\|_\infty) .$$

It follows that $\overline{P} = C([0,1])$. This completes the proof of the Weierstrass Approximation Theorem. $\square$

In relation to completion, Theorem 2.2.1 can be reformulated as follows.

Theorem 2.2.3. *The normed vector space $(C([a,b]), \|\cdot\|_\infty)$ is the completion of the normed vector space $(P, \|\cdot\|_\infty)$.*

2.3 Important inequalities for p-norms in spaces of continuous functions

Definition 2.3.1. The *support* of a function $f : \mathbb{R} \to \mathbb{C}$ is the closure of the set of points $x \in \mathbb{R}$ where $f(x) \neq 0$, i.e.

$$\text{support}(f) = \overline{\{x \in \mathbb{R} \mid f(x) \neq 0\}} \ .$$

If $\text{support}(f)$ is compact, we say that f has *compact support*.

Denote by $C_0(\mathbb{R})$ the set of continuous functions $f : \mathbb{R} \to \mathbb{C}$ with compact support.

It is easy to prove that $C_0(\mathbb{R})$ is a complex vector space with the obvious pointwise defined operations.

We shall now equip $C_0(\mathbb{R})$ with the so-called *p-norms* for $p \geq 1$. In this context, p is not necessarily an integer.

Definition 2.3.2. For an arbitrary $p \geq 1$, we define

$$||\cdot||_p : C_0(\mathbb{R}) \to \mathbb{R} \quad \text{by} \quad ||f||_p = \left(\int_{\mathbb{R}} |f(x)|^p dx \right)^{1/p} ,$$

for all functions $f : \mathbb{R} \to \mathbb{C}$ in $C_0(\mathbb{R})$.

Remark 2.3.3. Note that there are no problems in the above with the existence of the (Riemann) integral, since f is continuous and has compact support.

If $\text{support}(f) \subseteq [a, b]$, we have in fact

$$\int_{\mathbb{R}} |f(x)|^p dx = \int_a^b |f(x)|^p dx \ .$$

All constructions and conclusions in the following will also make sense for the vector space $C([a, b])$ of continuous functions $f : [a, b] \to \mathbb{C}$ defined in an arbitrary closed and bounded interval $[a, b]$.

Theorem 2.3.4. *The function $||\cdot||_p : C_0(\mathbb{R}) \to \mathbb{R}$, $p \geq 1$, is a norm in $C_0(\mathbb{R})$, known as the p-norm in $C_0(\mathbb{R})$.*

Proof. NORM 1: $||f||_p \geq 0$, $= 0 \Longleftrightarrow f = 0$ is clear by continuity of f. NORM 2 follows by the computation

$$(||\alpha f||_p)^p = \int_{\mathbb{R}} |(\alpha f)(x)|^p dx = |\alpha|^p \int_{\mathbb{R}} |f(x)|^p dx = \big(|\alpha| \, ||f||_p\big)^p \ .$$

The problem is NORM 3: the triangle inequality. The triangle inequality for the p-norm is known as the Minkowski inequality, which we shall prove below in Theorem 2.3.8. □

For the proof of Minkowski's inequality we need some inequalities of independent interest.

Theorem 2.3.5 (Young's inequality). *Let $p > 1$, $q > 1$ be a pair of real numbers satisfying $1/p + 1/q = 1$ (conjugate numbers). Then the following inequality holds for any pair of non-negative real numbers $a \geq 0$, $b \geq 0$,*

$$ab \leq \frac{1}{p}a^p + \frac{1}{q}b^q \ .$$

Proof. Consider the graph of the function $y = x^{p-1}$, which is also the graph of the function $x = y^{1/(p-1)} = y^{q-1}$, since

$$p - 1 = \frac{p}{q} \quad \text{and} \quad q - 1 = \frac{q}{p} \ .$$

By comparison of areas in Figure 2.2, we get

$$ab \leq \text{area}(I) + \text{area}(II) = \int_0^a x^{p-1}dx + \int_0^b y^{q-1}dy = \frac{a^p}{p} + \frac{b^q}{q} \ .$$
□

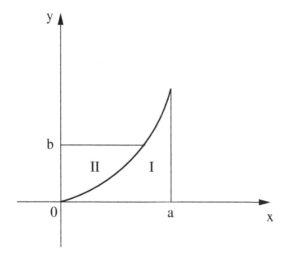

Fig. 2.2   Young's inequality

The following inequality established by the German mathematician Otto
Hölder (1859–1937) is the fundamental key to most of the important inequalities involving norms in function spaces.

Theorem 2.3.6 (Hölder's inequality). *Let $p > 1$, $q > 1$ be a pair of real
numbers satisfying $1/p + 1/q = 1$. For any pair of continuous functions $f, g \in
C_0(\mathbb{R})$, the following inequality holds*

$$\|fg\|_1 \leq \|f\|_p \|g\|_q .$$

Proof.   The inequality is clearly satisfied if either f or g is the constant
function with value 0. Hence we can assume that $\|f\|_p > 0$ and $\|g\|_q > 0$.
   By Young's inequality we first get,

$$\frac{|f(x)g(x)|}{\|f\|_p \|g\|_q} = \frac{|f(x)|}{\|f\|_p} \frac{|g(x)|}{\|g\|_q} \leq \frac{1}{p} \frac{|f(x)|^p}{\|f\|_p^{\,p}} + \frac{1}{q} \frac{|g(x)|^q}{\|g\|_q^{\,q}} ,$$

and then by integration,

$$\int_{\mathbb{R}} \frac{|f(x)g(x)|}{\|f\|_p \|g\|_q} dx \leq \frac{1}{p} \int_{\mathbb{R}} \frac{|f(x)|^p}{\|f\|_p^{\,p}} dx + \frac{1}{q} \int_{\mathbb{R}} \frac{|g(x)|^q}{\|g\|_q^{\,q}} dx$$
$$= \frac{1}{p} \frac{\|f\|_p^{\,p}}{\|f\|_p^{\,p}} + \frac{1}{q} \frac{\|g\|_q^{\,q}}{\|g\|_q^{\,q}} = \frac{1}{p} + \frac{1}{q} = 1 ,$$

from which follows that $\|fg\|_1 \leq \|f\|_p \|g\|_q$. $\square$

For $p = q = 2$, we get the Cauchy-Schwarz inequality.

Corollary 2.3.7 (Cauchy-Schwarz' inequality). *For any pair of continuous
functions $f, g \in C_0(\mathbb{R})$, the following inequality holds*

$$\|fg\|_1 \leq \|f\|_2 \|g\|_2 .$$

In detail:

$$\left| \int_{\mathbb{R}} f(x)g(x)dx \right| \leq \int_{\mathbb{R}} |f(x)g(x)|dx \leq \left(\int_{\mathbb{R}} f(x)^2 dx \right)^{1/2} \left(\int_{\mathbb{R}} g(x)^2 dx \right)^{1/2} .$$

   Now we can prove the triangle inequality for the p-norm. The inequality is
due to the Russian-German mathematician Hermann Minkowski (1864–1909).

Theorem 2.3.8 (Minkowski's inequality). *For any real number $p \geq 1$, and
any pair of continuous functions $f, g \in C_0(\mathbb{R})$, the following inequality holds*

$$\|f + g\|_p \leq \|f\|_p + \|g\|_p .$$

Proof. For $p = 1$, the proof follows by an easy computation:

$$||f + g||_1 = \int_{\mathbb{R}} |f(x) + g(x)| dx \leq \int_{\mathbb{R}} |f(x)| dx + \int_{\mathbb{R}} |g(x)| dx = ||f||_1 + ||g||_1 .$$

For $p > 1$, put $1/q = 1 - 1/p$. Then we have a pair of real numbers $p > 1$, $q > 1$, satisfying $1/p + 1/q = 1$.

First an elementary rewriting:

$$I = \int_{\mathbb{R}} |f(x) + g(x)|^p dx = \int_{\mathbb{R}} |f(x) + g(x)| \, |f(x) + g(x)|^{p-1} dx .$$

Applying the triangle inequality for numerical values, we then get

$$I \leq \int_{\mathbb{R}} |f(x)| \, |f(x) + g(x)|^{p-1} dx + \int_{\mathbb{R}} |g(x)| \, |f(x) + g(x)|^{p-1} dx ,$$

and by rewriting to 1-norms and applying Hölder's inequality,

$$\int_{\mathbb{R}} |f(x) + g(x)|^p dx \leq ||f \, (f+g)^{p-1}||_1 + ||g \, (f+g)^{p-1}||_1$$

$$\leq ||f||_p \, ||(f+g)^{p-1}||_q + ||g||_p \, ||(f+g)^{p-1}||_q$$

$$= (||f||_p + ||g||_p)\left(\int_{\mathbb{R}} |f(x) + g(x)|^{(p-1)q} dx \right)^{1/q}$$

$$= (||f||_p + ||g||_p)\left(\int_{\mathbb{R}} |f(x) + g(x)|^p dx \right)^{1-1/p} .$$

If $\int_{\mathbb{R}} |f(x) + g(x)|^p dx = 0$, we have $||f + g||_p = 0$, and the inequality is trivially true.

If $\int_{\mathbb{R}} |f(x) + g(x)|^p dx \neq 0$, we can divide by $\left(\int_{\mathbb{R}} |f(x) + g(x)|^p dx \right)^{1-1/p}$ and get to the inequality

$$\left(\int_{\mathbb{R}} |f(x) + g(x)|^p dx \right)^{1-(1-1/p)} \leq ||f||_p + ||g||_p ,$$

which reduces to

$$||f + g||_p \leq ||f||_p + ||g||_p .$$

This completes the proof. $\square$

2.4 Construction of L^p-spaces

The theory of L^p-spaces plays a fundamental role in applications of functional analysis, for example in the theory of differential and integral equations. It is a delicate subject involving a considerable amount of mathematical abstraction.

The notation 'L^p-space' signals a normed vector space of 'p^{th} power (absolute) Lebesgue integrable functions' on a measure space. The name honors the French mathematician Henri Lebesgue (1875–1941) who introduced a more general notion of integral than the Riemann integral, which allows many more functions in an interval to be integrable than the continuous functions. The vector space of all the pth power Lebesgue integrable functions is complete in the p-norm, in other words, it is a Banach space in the p-norm.

In this book we use an approach to the construction of L^p-spaces on the real line avoiding the theory of Lebesgue integration. We construct the L^p-spaces directly using the completion process developed in Theorem 2.1.6.

2.4.1 *The L^p-spaces and some basic inequalities*

Definition 2.4.1. The vector space $L^p(\mathbb{R})$, $p \geq 1$, is the completion of $C_0(\mathbb{R})$ with respect to the p-norm in $C_0(\mathbb{R})$.

The norm in $L^p(\mathbb{R})$ induced by the p-norm in $C_0(\mathbb{R})$ is also denoted by $||\cdot||_p$. The pair $(L^p(\mathbb{R}), ||\cdot||_p)$ is then a Banach space.

Remark 2.4.2. By construction, the vector space $C_0(\mathbb{R})$ is a subspace of the Banach space $L^p(\mathbb{R})$ for all $p \geq 1$. But even more, $C_0(\mathbb{R})$ is a *dense* subspace of $L^p(\mathbb{R})$ for all $p \geq 1$.

Remark 2.4.3. It is not necessary to know the exact nature of the elements in $L^p(\mathbb{R})$. It is sufficient to work with them as limits of Cauchy sequences of functions in $C_0(\mathbb{R})$ with respect to the p-norm $||\cdot||_p$. The elements of $L^p(\mathbb{R})$ are called L^p-*functions*, or *functions of class L^p*, and we use notation as that known from the theory of functions. An L^p-function f is then a (possibly added) limit of a Cauchy sequence (f_n) of ordinary continuous functions $f_n : \mathbb{R} \to \mathbb{C}$ with compact support. In other words, every L^p-function $f : \mathbb{R} \to \mathbb{C}$ is the limit (in the p-norm) of a Cauchy sequence (f_n) in $(C_0(\mathbb{R}), ||\cdot||_p)$, i.e. $f_n \to f$ for $n \to \infty$.

Remark 2.4.4. Some of the L^p-functions are ordinary continuous functions, in fact, a dense subset of them are continuous functions, namely the functions in $C_0(\mathbb{R})$. But there are many more L^p-functions than the continuous functions; exactly as in the relation between the set of real numbers $\mathbb{R}$ and the set of rational numbers $\mathbb{Q}$. The completion entails among others that if we have a function $f \in C_0(\mathbb{R})$ and change it arbitrarily in a countable set of points in $\mathbb{R}$, or, more generally in a set with Lebesgue measure zero (cf. Definition 2.4.11), then we get a function of class L^p; in this particular case the new function is actually identical to f in $L^p(\mathbb{R})$.

Remark 2.4.5. In addition to the L^p-spaces, $p \geq 1$, we also have the space $L^\infty(\mathbb{R})$. This space is more delicate to define. It can be constructed by a completion process using a variant of the uniform norm from the space of complex-valued functions in $\mathbb{R}$ that are of class L^1 in every bounded interval of $\mathbb{R}$ and are *essentially bounded*, i.e. bounded except in a set of Lebesgue measure zero.

If we consider the space of continuous functions $C([a,b])$ in a fixed closed and bounded interval $[a,b]$, Proposition 2.1.2 shows that $C([a,b])$ is complete in the uniform norm. Hence $C([a,b])$ is a closed subspace of $L^\infty([a,b])$.

The vector space structure in the normed vector space $(C_0(\mathbb{R}), ||\cdot||_p)$ can be enriched with a further structure, namely a multiplication. For any pair of functions $f, g \in C_0(\mathbb{R})$, the function $fg \in C_0(\mathbb{R})$, given by the obvious pointwise defined operation, is again a function in $C_0(\mathbb{R})$, just as in the case of the functions $f + g$ and αf for an arbitrary scalar $\alpha \in \mathbb{C}$. Altogether, $C_0(\mathbb{R})$ has algebraic structure as a so-called *function algebra*.

For the Banach spaces $(L^p(\mathbb{R}), ||\cdot||_p)$, $p \geq 1$, the situation is more complicated with respect to multiplication; the L^p-class may drop under the multiplication. We do, however, have the following result.

Lemma 2.4.6. *Let $p > 1$, $q > 1$ be a conjugate pair of real numbers, i.e. $1/p + 1/q = 1$. For any pair of functions $f \in L^p(\mathbb{R})$, $g \in L^q(\mathbb{R})$, we can then define a product function $fg \in L^1(\mathbb{R})$.*

Proof. Suppose that $f = [(f_n)]$ for a Cauchy sequence (f_n) in $(C_0(\mathbb{R}), ||\cdot||_p)$, and that $g = [(g_n)]$ for a Cauchy sequence (g_n) in $(C_0(\mathbb{R}), ||\cdot||_q)$. In the terminology introduced for L^p-spaces, we have that

$$f_n \to f \quad \text{for} \quad n \to \infty \quad \left(\text{in } (L^p(\mathbb{R}), ||\cdot||_p) \right)$$
$$g_n \to g \quad \text{for} \quad n \to \infty \quad \left(\text{in } (L^q(\mathbb{R}), ||\cdot||_q) \right).$$

Assertion 2.4.7. *The sequence of continuous functions $(f_n g_n)$ is a Cauchy sequence in $(C_0(\mathbb{R}), ||\cdot||_1)$.*

Proof. For Cauchy sequences, the norms are bounded, and hence we can choose positive real constants k_1, k_2 such that

$$||f_n||_p \leq k_1 \text{ and } ||g_n||_q \leq k_2 \text{ for all } n \geq 1 .$$

Making use first of the triangle inequality and next of Hölder's inequality for continuous functions, Theorem 2.3.6, we get the following estimates,

$$
\begin{aligned}
\|f_n g_n - f_m g_m\|_1 &= \|(f_n - f_m)g_n + f_m(g_n - g_m)\|_1 \\
&\leq \|(f_n - f_m)g_n\|_1 + \|f_m(g_n - g_m)\|_1 \\
&\leq \|(f_n - f_m)\|_p \|g_n\|_q + \|f_m\|_p \|(g_n - g_m)\|_q \\
&\leq \|(f_n - f_m)\|_p k_2 + k_1 \|(g_n - g_m)\|_q \ ,
\end{aligned}
$$

from which follows that $(f_n g_n)$ is a Cauchy sequence in $(C_0(\mathbb{R}), \|\cdot\|_1)$. $\qquad\square$

Since $(f_n g_n)$ is a Cauchy sequence in $(C_0(\mathbb{R}), \|\cdot\|_1)$, it defines a function of class L^1. For obvious reasons we denote this function by fg. With this definition of fg we have

$$
f_n g_n \to fg \quad \text{for} \quad n \to \infty \quad \text{in} \ (L^1(\mathbb{R}), \|\cdot\|_1) \ .
$$

This completes the proof of Lemma 2.4.6 including the definition of product functions for certain pairs of L^p-functions. $\qquad\square$

The inequalities of Hölder, Cauchy-Schwarz and Minkowski can be extended to L^p-spaces by limit arguments in which L^p-functions are represented by Cauchy sequences of continuous functions with compact support. We get the following results.

Theorem 2.4.8 (Hölder's inequality). *Let $p > 1$, $q > 1$ be a pair of real numbers satisfying $1/p + 1/q = 1$. For any pair of functions $f \in L^p(\mathbb{R})$, $g \in L^q(\mathbb{R})$, the product function $fg \subset L^1(\mathbb{R})$, and the following inequality holds*

$$
\|fg\|_1 \leq \|f\|_p \|g\|_q \ .
$$

Corollary 2.4.9 (Cauchy-Schwarz' inequality). *For any pair of L^2-functions $f, g \in L^2(\mathbb{R})$, the product function $fg \in L^1(\mathbb{R})$, and the following inequality holds*

$$
\|fg\|_1 \leq \|f\|_2 \|g\|_2 \ .
$$

Theorem 2.4.10 (Minkowski's inequality). *For any real number $p \geq 1$, and any pair of L^p-functions $f, g \in L^p(\mathbb{R})$, the following inequality holds*

$$
\|f + g\|_p \leq \|f\|_p + \|g\|_p \ .
$$

2.4.2 *Lebesgue measurable subsets in* $\mathbb{R}$

The notion of L^1-functions can be used to introduce the notion of *Lebesgue measure* in $\mathbb{R}$. We present a few details in such an approach.

For any subset $A \subseteq \mathbb{R}$ of the real line, define the *indicator function* f_A for A by

$$f_A(x) = \begin{cases} 1 & \text{for } x \in A \\ 0 & \text{otherwise.} \end{cases}$$

Definition 2.4.11. A subset $A \subseteq \mathbb{R}$ is said to be *Lebesgue measurable* with *finite measure* if the indicator function f_A for A is of class L^1, i.e. if $f_A \in L^1(\mathbb{R})$.

More generally, a subset $A \subseteq \mathbb{R}$ is said to be *Lebesgue measurable* if the intersection of A and the closed interval $[-n, n]$ is Lebesgue measurable with finite measure for each $n \in \mathbb{N}$.

In case A is Lebesgue measurable with finite measure, the *Lebesgue measure* of A is the number $\mu(A)$ given by

$$\mu(A) = ||f_A||_1 = \int_{\mathbb{R}} f_A(x)dx \ .$$

In the general case, the *Lebesgue measure* of a Lebesgue measurable set A is the limit

$$\mu(A) = \lim_{n \to \infty} \mu\big(A \cap [-n, n]\big) \quad (\text{possibly } \infty) \ .$$

Example 2.4.12. Let A be an arbitrary bounded interval in $\mathbb{R}$ with endpoints $a < b$. The interval can be open, closed, or half open as appropriate.

In the spirit of Example 2.1.4, we can construct a Cauchy sequence (f_n) in $(C_0(\mathbb{R}), ||\cdot||_1)$ such that $f_n \to f_A$ for $n \to \infty$ in $(L^1(\mathbb{R}), ||\cdot||_1)$, by defining

$$f_n(x) = \begin{cases} 0 & \text{for } x \le a - 1/n \\ n(x - a + 1/n) & \text{for } a - 1/n \le x \le a \\ 1 & \text{for } a \le x \le b \\ 1 - n(x - b) & \text{for } b \le x \le b + 1/n \\ 0 & \text{for } b + 1/n \le x \end{cases} \ .$$

This proves that $f_A \in L^1(\mathbb{R})$ and that the Lebesgue measure of the interval A is given by

$$\mu(A) = ||f_A||_1 = \lim_{n \to \infty} ||f_n||_1 = \int_a^b 1\, dx = b - a \ .$$

The system of bounded intervals in $\mathbb{R}$ has a number of useful properties with respect to Lebesgue measure.

Theorem 2.4.13.

(1) *An arbitrary bounded interval I in $\mathbb{R}$ is Lebesgue measurable and the Lebesgue measure $\mu(I)$ of I is the length of the interval.*

(2) *An arbitrary finite union of mutually disjoint, bounded intervals $I_1, \ldots, I_n$ is Lebesgue measurable and*

$$\mu\left(\bigcup_{k=1}^{n} I_k\right) = \sum_{k=1}^{n} \mu(I_k) .$$

(3) *For any two bounded intervals I and J in $\mathbb{R}$ it holds that*

$$\mu(I \cup J) + \mu(I \cap J) = \mu(I) + \mu(J) .$$

(4) *Let $\{I_n \,|\, n \in \mathbb{N}\}$ be a countable collection of bounded intervals for which the infinite series $\sum_{n=1}^{\infty} \mu(I_n)$ is convergent. Then the union $\bigcup_{n=1}^{\infty} I_n$ is Lebesgue measurable and it holds that (equality for mutually disjoint intervals)*

$$\mu\left(\bigcup_{n=1}^{\infty} I_n\right) \leq \sum_{n=1}^{\infty} \mu(I_n) .$$

(5) *A subset A in $\mathbb{R}$ is Lebesgue measurable with $\mu(A) = 0$ if for every $\varepsilon > 0$, there exists a countable union of open intervals $O = \bigcup_{n=1}^{\infty} I_n$ in $\mathbb{R}$ with Lebesgue measure $\mu(O) < \varepsilon$ such that $A \subseteq O$.*

Proof.　　The result in (1) is proved in Example 2.4.12.

Under the assumptions in (2), the indicator function for $\bigcup_{k=1}^{n} I_k$ is the sum of the indicator functions for I_k, $k = 1, \ldots, n$, which proves (2).

(3) follows by applying the result in (2) to the splitting

$$I \cup J = (I \setminus I \cap J) \cup (I \cap J) \cup (J \setminus I \cap J) .$$

For the proof of (4) define subsets J_n of $\mathbb{R}$ by setting $J_n = I_n \setminus \bigcup_{k=1}^{n-1} I_k$, for $n \geq 2$, and $J_1 = I_1$. Then the subsets $\{J_n \,|\, n \in \mathbb{N}\}$ are mutually disjoint and $A = \bigcup_{n=1}^{\infty} I_n = \bigcup_{n=1}^{\infty} J_n$. Each J_n is Lebesgue measurable with $\mu(J_n) \leq \mu(I_n)$, since it is a finite union of subintervals of I_n. The indicator function for A can be written as the sum $f_A = \sum_{n=1}^{\infty} f_{J_n}$, which proves that $f_A \in L^1(\mathbb{R})$, and hence that A is Lebesgue measurable. Furthermore, $\mu(A) = \mu(\bigcup_{n=1}^{\infty} I_n) = \mu(\bigcup_{n=1}^{\infty} J_n) \leq \sum_{n=1}^{\infty} \mu(I_n)$, which proves (4).

(5) follows since $A \subseteq \bigcap_{n=1}^{\infty} O_n$ for open sets O_n with $\mu(O_n) < 1/n$.　　□

Example 2.4.14. Let $A = \{r_1, r_2, \ldots, r_n, \ldots\}$ be a countable subset of $\mathbb{R}$. We shall prove that $\mu(A) = 0$. With this in mind, let $\varepsilon > 0$ be arbitrarily given. Then A can be enclosed in an open subset $O \subseteq \mathbb{R}$ with Lebesgue measure

$\mu(O) < \varepsilon$ by choosing an open interval I_n of length $\varepsilon/2^n$ containing r_n for all $n \in \mathbb{N}$ and letting $O = \bigcup_{n=1}^{\infty} I_n$. By property (5) in Theorem 2.4.13 it follows that A is Lebesgue measurable with $\mu(A) = 0$.

Example 2.4.15. We shall construct the *Cantor set C*, introduced by and named after the German mathematician Georg Cantor (1845–1918).

In the first step of the construction, we subdivide the unit interval $[0, 1]$ into three subintervals of equal lengths and then exclude the open middle interval $(\frac{1}{3}, \frac{2}{3})$. In the second step, we divide each of the two intervals $[0, \frac{1}{3}]$ and $[\frac{2}{3}, 1]$ into three equal parts and exclude $(\frac{1}{9}, \frac{2}{9})$, respectively $(\frac{7}{9}, \frac{8}{9})$, from those. Continue this way ad infinitum, in step n dividing each of the remaining closed intervals from step $n-1$ into three equal parts and excluding the open middle intervals. In the limit we get the *Cantor set C*.

We can view C as the complement $C = [0, 1] \setminus S$, where S is the union of the recursively defined open sets,

$$S_1 = \left(\frac{1}{3}, \frac{2}{3}\right)$$

$$S_2 = \left(\frac{1}{9}, \frac{2}{9}\right) \cup \left(\frac{7}{9}, \frac{8}{9}\right)$$

$$S_3 = \left(\frac{1}{27}, \frac{2}{27}\right) \cup \left(\frac{7}{27}, \frac{8}{27}\right) \cup \left(\frac{19}{27}, \frac{20}{27}\right) \cup \left(\frac{25}{27}, \frac{26}{27}\right)$$

$$\vdots$$

Arguing as in Example 2.4.12, it can be proved that S is Lebesgue measurable and with Lebesgue measure,

$$\mu(S) = \int_0^1 f_S(x)dx = \int_0^1 1\,dx = 1 \ .$$

Since $f_C = 1 - f_S$, it follows that $f_C \in L^1(\mathbb{R})$, and hence that the Cantor set C is Lebesgue measurable. We also get that the Cantor set has Lebesgue measure 0, since

$$\mu(C) = \int_{\mathbb{R}} f_C(x)dx = \int_0^1 (1 - f_S(x))dx = 1 - 1 = 0 \ .$$

If we write the numbers in $[0, 1]$ as power series $\sum_1^{\infty} a_n(1/3)^n$, with $a_n = 0, 1, 2$ (triadic expansion), then the numbers in C are represented by triadic expansions with coefficients a_n either 0 or 2. We can then prove that the set C is uncountable by an indirect argument. For suppose $r_1, r_2, \ldots, r_n, \ldots$ is the complete list of numbers in C. Then construct a number r with triadic expansion $r = \sum_1^{\infty} a_n(1/3)^n$ by choosing the coefficient a_n to be either 0 or

2 but different from the nth coefficient in r_n for each $n \in \mathbb{N}$. Now r is a new number in C contradicting that $r_1, r_2, \ldots, r_n, \ldots$ is the complete list of numbers in C.

The Cantor set C provides an example of an uncountable subset of $[0,1]$ with Lebesgue measure 0, thus proving that Lebesgue measure 0 is not restricted to countable sets as in Example 2.4.14.

2.4.3 *Smooth functions with compact support*

There are other nice subspaces that are dense in the L^p-spaces than the space $C_0(\mathbb{R})$ of continuous functions with compact support.

Definition 2.4.16. Denote by $C_0^\infty(\mathbb{R})$ the set of functions $f : \mathbb{R} \to \mathbb{C}$ that are infinitely often differentiable (smooth) and have compact support.

Example 2.4.17. One may wonder if such functions do exist. Here is one with compact support in the closed and bounded interval $[a, b]$:

$$\varphi_{a,b}(x) = \begin{cases} \exp\left(\frac{-1}{(x-a)(b-x)}\right) & \text{for } x \in {]a,b[} \\ 0 & \text{otherwise.} \end{cases}$$

Differentiability at $x = a$ and $x = b$ is ensured by the fact that exponential functions grow faster than polynomial functions. For good reasons, the function $\varphi_{a,b}$ is called a *bump function*.

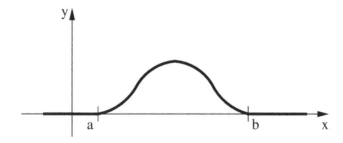

Fig. 2.3   Graph of $\varphi_{a,b}$

We state the following useful result without proof.

Theorem 2.4.18.

(1) *The space $C_0^\infty(\mathbb{R})$ is dense in $C_0(\mathbb{R})$ both with respect to the uniform norm (supremum norm) and to the p-norm, $p \geq 1$.*

(2) *The space $C_0^\infty(\mathbb{R})$ is dense in all the spaces $L^p(\mathbb{R})$, $p \geq 1$.*

2.4.4   Riemann integrable functions

The space of Riemann integrable functions is the classical space of functions considered in connection with integration theory. In this section we shall demonstrate that the space of absolute p^{th} power Riemann integrable functions can be identified with a dense subspace of the corresponding L^p-space.

First we recall the definition of Riemann integrability.

Let $f : [a, b] \to \mathbb{C}$ be a bounded function defined in a closed and bounded interval $[a, b]$ in $\mathbb{R}$. For any *partition* $\mathcal{D}$ of the interval $[a, b]$ into a finite number of subintervals $[t_{i-1}, t_i]$, $1 \leq i \leq n$, marked by the points

$$\mathcal{D}: \quad a = t_0 < t_1 < \cdots < t_{n-1} < t_n = b \ ,$$

and for any selection $\tau_1, \tau_2, \ldots, \tau_n$ of points in $[a, b]$ such that $\tau_i \in [t_{i-1}, t_i]$ for $1 \leq i \leq n$, define the *Riemann sum*

$$S(f, \mathcal{D}) = \sum_{i=1}^{n} f(\tau_i)(t_i - t_{i-1}) \ .$$

The *norm* of the partition is the number

$$\mu(\mathcal{D}) = \max \left\{ |t_i - t_{i-1}| \,\big|\, i = 1, \ldots, n \right\} \ .$$

The function $f : [a, b] \to \mathbb{C}$ is said to be *Riemann integrable* if the collection of Riemann sums $S(f, \mathcal{D})$ has a limit for $\mu(\mathcal{D}) \to 0$. In case of integrability, the limit is called the *Riemann integral* of f and it is denoted by $\int_a^b f(x)dx$. In short, the Riemann integral is given by

$$\int_a^b f(x)dx = \lim_{\mu(\mathcal{D}) \to 0} S(f, \mathcal{D}) \ .$$

Now consider a bounded function $f : \mathbb{R} \to \mathbb{C}$ with compact support. By choosing a closed and bounded interval containing the support of f we can define the notion of Riemann integrability also for such functions. Let $\mathcal{R}_0(\mathbb{R})$ denote the space of bounded, Riemann integrable functions with compact support. We can introduce a vector space structure in $\mathcal{R}_0(\mathbb{R})$ by the obvious pointwise defined operations. Then $C_0(\mathbb{R})$ is a linear subspace in $\mathcal{R}_0(\mathbb{R})$.

For any real number $p \geq 1$, a bounded function $f : \mathbb{R} \to \mathbb{C}$ with compact support is said to be absolute p^{th} power Riemann integrable if the following integral exists in the sense of Riemann,

$$\int_{\mathbb{R}} |f(x)|^p dx .$$

Denote by $\mathcal{R}_0^p(\mathbb{R})$ the space of all such functions.

For every function $f \in \mathcal{R}_0^p(\mathbb{R})$, we can define the p-norm

$$|| f ||_p = \Big(\int_{\mathbb{R}} |f(x)|^p dx \Big)^{1/p} .$$

The inequalities of Hölder, Cauchy-Schwarz and Minkowski can be proved for the spaces $\mathcal{R}_0^p(\mathbb{R})$ by limit arguments with Riemann sums. In particular, we get the (semi) normed vector space $(\mathcal{R}_0^p(\mathbb{R}), || \cdot ||_p)$ containing the normed vector space $(C_0(\mathbb{R}), || \cdot ||_p)$ as a dense subspace.

Remark 2.4.19. The p-norm $|| \cdot ||_p$ is only a so-called *seminorm* in $\mathcal{R}_0^p(\mathbb{R})$ since it can happen that $|| f ||_p = 0$ for a nonzero $f \in \mathcal{R}_0^p(\mathbb{R})$. This will be of no importance for the validity of the proof of Theorem 2.4.20.

The pth power Riemann integrable functions with compact support can be considered as L^p-functions. This is the message of the following theorem.

Theorem 2.4.20. *For any $p \geq 1$, the space $(\mathcal{R}_0^p(\mathbb{R}), || \cdot ||_p)$ can be identified with a dense subspace in $(L^p(\mathbb{R}), || \cdot ||_p)$. With this identification, we have the linear and isometric (with respect to p-norm) inclusions*

$$C_0(\mathbb{R}) \subset \mathcal{R}_0^p(\mathbb{R}) \subset L^p(\mathbb{R})$$

and

$$\overline{C_0(\mathbb{R})} = \overline{\mathcal{R}_0^p(\mathbb{R})} = L^p(\mathbb{R}) .$$

Proof. The only thing to be proved is that a Cauchy sequence in $(\mathcal{R}_0^p(\mathbb{R}), || \cdot ||_p)$ is equivalent to a Cauchy sequence in $(C_0(\mathbb{R}), || \cdot ||_p)$. This follows by observing that if (f_n) is a Cauchy sequence in $(\mathcal{R}_0^p(\mathbb{R}), || \cdot ||_p)$, then we can construct a Cauchy sequence (g_n) in $(C_0(\mathbb{R}), || \cdot ||_p)$ such that $|| f_n - g_n ||_p \to 0$ for $n \to \infty$. This is left to the reader as an exercise. $\square$

We finish this section by stating (without proof) a result that characterizes the Riemann integrable functions among the Lebesgue integrable functions.

Theorem 2.4.21. *Let $f : [a, b] \to \mathbb{R}$ be a bounded, real-valued function defined in the closed and bounded interval $[a, b]$. Suppose that f is Lebesgue integrable,* *i.e.* $f \in L^1([a,b])$. *Then* $f : [a,b] \to \mathbb{R}$ *is a Riemann integrable function if and only if the set of points where* f *is not continuous is contained in a set of Lebesgue measure zero.*

Example 2.4.22. Let A be the subset of rational numbers in the closed unit interval $[0,1]$ and let $f_A : [0,1] \to \mathbb{R}$ be the indicator function for A. Since the set of rational numbers is countable, the function f_A is Lebesgue integrable with Lebesgue integral $\|f_A\|_1 = \int_0^1 f_A(x)\,dx = 0$.

On the other hand, the function f_A is not Riemann integrable since its Riemann sums fluctuate between 0 and 1. Note that there is no conflict with Theorem 2.4.21, even though the Lebesgue measure $\mu(A) = 0$, since f_A is not continuous at any point of $[0,1]$.

2.5 The sequence spaces l^p

Corresponding to the function spaces $L^p(\mathbb{R})$, there are similar notions for spaces of sequences (x_n) in $\mathbb{R}$ (or $\mathbb{C}$).

This is not surprising since a sequence (x_n) can actually be viewed as a function $f : \mathbb{N} \to \mathbb{R}$ (or $\mathbb{C}$) by the identification $x_n = f(n)$, $n \in \mathbb{N}$. To transfer results from the continuous case to the discrete case, we just have to substitute $\int_{\mathbb{R}} dx$ by $\sum_{n=1}^{\infty}$; for example $\int_{\mathbb{R}} |f(x)|^p dx$ should be substituted by $\sum_{n=1}^{\infty} |x_n|^p$.

This leads to the so-called l^p-spaces, $p \geq 1$.

Definition 2.5.1. For any real number $p \geq 1$, a (real or complex) sequence (x_n) is said to be *absolute p-summable*, if the infinite series of pth-powers of absolute values of the elements in the sequence is convergent, i.e.

$$\sum_{n=1}^{\infty} |x_n|^p < \infty .$$

(i) Denote by l^p, $p \geq 1$, the space of all absolute p-summable sequences. The norm in l^p is defined by

$$\|(x_n)\|_p = \left(\sum_{n=1}^{\infty} |x_n|^p \right)^{1/p} .$$

(ii) Denote by l^∞ the sequence space of all bounded sequences (x_n). The norm in l^∞ is defined by

$$\|(x_n)\|_\infty = \sup_{n \in \mathbb{N}} |x_n| .$$

To justify the terminology used in Definition 2.5.1, we need to prove that l^p is indeed a normed vector space with the norm $||\cdot||_p$. This will follow by Minkowski's inequality for sequences, which we shall prove below.

First we prove Hölder's inequality for sequences.

Theorem 2.5.2 (Hölder's inequality). *Let $p > 1$, $q > 1$ be a pair of real numbers satisfying $1/p + 1/q = 1$. For any pair of sequences $(x_n) \in l^p$, $(y_n) \in l^q$, the product sequence $(x_n y_n) \in l^1$, and the following inequality holds*

$$|| (x_n y_n) ||_1 \leq || (x_n) ||_p || (y_n) ||_q .$$

In detail:

$$\sum_{n=1}^{\infty} |x_n y_n| \leq \Big(\sum_{n=1}^{\infty} |x_n|^p \Big)^{1/p} \Big(\sum_{n=1}^{\infty} |y_n|^q \Big)^{1/q} .$$

Proof.   The inequality is clearly satisfied if either (x_n) or (y_n) is the constant sequence with all elements 0. Hence we can assume that $|| (x_n) ||_p > 0$ and $|| (y_n) ||_q > 0$.

By Young's inequality, Theorem 2.3.5, we first get,

$$\frac{|x_n y_n|}{|| (x_n) ||_p || (y_n) ||_q} = \frac{|x_n|}{|| (x_n) ||_p} \frac{|y_n|}{|| (y_n) ||_q} \leq \frac{1}{p} \frac{|x_n|^p}{|| (x_n) ||_p^p} + \frac{1}{q} \frac{|y_n|^q}{|| (y_n) ||_q^q} ,$$

and then by summation,

$$\sum_{n=1}^{N} \frac{|x_n y_n|}{|| (x_n) ||_p || (y_n) ||_q} \leq \frac{1}{p} \sum_{n=1}^{N} \frac{|x_n|^p}{|| (x_n) ||_p^p} + \frac{1}{q} \sum_{n=1}^{N} \frac{|y_n|^q}{|| (y_n) ||_q^q}$$

$$\leq \frac{1}{p} \frac{|| (x_n) ||_p^p}{|| (x_n) ||_p^p} + \frac{1}{q} \frac{|| (y_n) ||_q^q}{|| (y_n) ||_q^q} = \frac{1}{p} + \frac{1}{q} = 1 ,$$

from which follows that $|| (x_n y_n) ||_1 \leq || (x_n) ||_p || (y_n) ||_q$ by letting $N \to \infty$. $\square$

As a special case, we get Cauchy-Schwarz' inequality.

Corollary 2.5.3 (Cauchy-Schwarz' inequality). *For any pair of square-summable sequences $(x_n), (y_n) \in l^2$, the product sequence $(x_n y_n) \in l^1$, and the following inequality holds*

$$|| (x_n y_n) ||_1 \leq || (x_n) ||_2 || (y_n) ||_2 .$$

In detail:

$$\sum_{n=1}^{\infty} |x_n y_n| \leq \Big(\sum_{n=1}^{\infty} |x_n|^2 \Big)^{1/2} \Big(\sum_{n=1}^{\infty} |y_n|^2 \Big)^{1/2} .$$

As the following theorem will show, the sum of two sequences in l^p is again a sequence in l^p, thereby proving (the difficult part of the argument) that l^p is a vector space. Furthermore, the inequality in the theorem is exactly the triangle inequality for $||\cdot||_p$.

Theorem 2.5.4 (Minkowski's inequality). *For any real number $p \geq 1$, and any pair of p-summable sequences $(x_n), (y_n) \in l^p$, the sum sequence $(x_n + y_n)$ is p-summable, i.e. $(x_n + y_n) \in l^p$, and the following inequality holds*

$$|| (x_n + y_n) ||_p \leq || (x_n) ||_p + || (y_n) ||_p .$$

In detail:

$$\left(\sum_{n=1}^{\infty} | x_n + y_n |^p \right)^{1/p} \leq \left(\sum_{n=1}^{\infty} | x_n |^p \right)^{1/p} + \left(\sum_{n=1}^{\infty} | y_n |^p \right)^{1/p} .$$

Proof. For $p = 1$, the proof follows by using the triangle inequality for the numerical value to obtain the estimates

$$\sum_{n=1}^{N} | x_n + y_n | \leq \sum_{n=1}^{N} | x_n | + \sum_{n=1}^{N} | y_n | \leq \sum_{n=1}^{\infty} | x_n | + \sum_{n=1}^{\infty} | y_n | ,$$

and then letting $N \to \infty$.

For $p > 1$, put $1/q = 1 - 1/p$. Then we have a pair of real numbers $p > 1$, $q > 1$, satisfying $1/p + 1/q = 1$.

First an elementary rewriting,

$$S_N = \sum_{n=1}^{N} | x_n + y_n |^p = \sum_{n=1}^{N} | x_n + y_n | \, | x_n + y_n |^{p-1} .$$

Applying the triangle inequality for the numerical value, we then get

$$S_N \leq \sum_{n=1}^{N} | x_n | \, | x_n + y_n |^{p-1} + \sum_{n=1}^{N} | y_n | \, | x_n + y_n |^{p-1} .$$

Applying Hölder's inequality for 'finite' sequences (N-tuples) we next get

$$S_N \leq \left(\sum_{n=1}^{N} |x_n|^p \right)^{1/p} \left(\sum_{n=1}^{N} |x_n + y_n|^{(p-1)q} \right)^{1/q}$$

$$+ \left(\sum_{n=1}^{N} |y_n|^p \right)^{1/p} \left(\sum_{n=1}^{N} |x_n + y_n|^{(p-1)q} \right)^{1/q}$$

$$\leq ||(x_n)||_p \left(\sum_{n=1}^{N} |x_n + y_n|^p \right)^{1/q} + ||(y_n)||_p \left(\sum_{n=1}^{N} |x_n + y_n|^p \right)^{1/q}$$

$$= (||(x_n)||_p + ||(y_n)||_p)(S_N)^{1/q} = (||(x_n)||_p + ||(y_n)||_p)(S_N)^{1-1/p} .$$

If $S_N = 0$ for all $N \geq 1$, clearly $||(x_n + y_n)||_p = 0$, and in this case, Minkowski's inequality is trivially true.

If $S_N \neq 0$ for N sufficiently large, we can divide by $(S_N)^{1-1/p}$, and get to the inequality

$$(S_N)^{1-(1-1/p)} \leq ||(x_n)||_p + ||(y_n)||_p ,$$

which in the limit $N \to \infty$ reduces to

$$||(x_n + y_n)||_p \leq ||(x_n)||_p + ||(y_n)||_p .$$

This completes the proof.   □

It is now easy to finish the proof that all the sequence spaces are normed vector spaces. We leave it as an exercise to the reader to prove that they are actually complete normed vector spaces, including the special sequence space l^∞. Collecting all facts we have the following main result.

Theorem 2.5.5. *All the sequence spaces* $(l^p, ||\cdot||_p)$, $p \geq 1$, *and* $(l^\infty, ||\cdot||_\infty)$ *are Banach spaces.*

In the continuous case, $C_0(\mathbb{R})$ is dense in all the L^p-spaces. In the sequence spaces, we have to replace 'compact support' of the functions in $C_0(\mathbb{R})$ by 'only finitely many elements different from zero' in the sequences in l^p.

Theorem 2.5.6. *The space of sequences with only a finite number of elements different from zero is dense in* $(l^p, ||\cdot||_p)$, *for all* $p \geq 1$.

Chapter 3

Theory of Hilbert Spaces

3.1 Inner product spaces

Let V be a complex vector space.

Definition 3.1.1. An *inner product* in V is a mapping $(\cdot, \cdot) : V \times V \to \mathbb{C}$ satisfying

IP 1 (Hermitian symmetry)
$$(x, y) = \overline{(y, x)} \quad \text{all } x, y \in V.$$

IP 2 (linearity)
$$(\alpha x_1 + \beta x_2, y) = \alpha(x_1, y) + \beta(x_2, y) \quad \text{all } x_1, x_2, y \in V, \alpha, \beta \in \mathbb{C}.$$

IP 3 (positive definite)
$$(x, x) \geq 0 \quad \text{and} \quad (x, x) = 0 \Leftrightarrow x = 0 \quad \text{all } x \in V.$$

Remark 3.1.2. The condition IP 1 implies that (x, x) is a real number such that condition IP 3 makes sense.

Remark 3.1.3. Exploiting conditions IP 1 and IP 2 we get,

$$
\begin{aligned}
(x, \alpha y_1 + \beta y_2) &= \overline{(\alpha y_1 + \beta y_2, x)} \\
&= \overline{\alpha(y_1, x) + \beta(y_2, x)} \\
&= \overline{\alpha}\,\overline{(y_1, x)} + \overline{\beta}\,\overline{(y_2, x)} \\
&= \overline{\alpha}\,(x, y_1) + \overline{\beta}\,(x, y_2) .
\end{aligned}
$$

By IP 2, the inner product is linear in the first argument. The above computation shows that the inner product is *conjugate linear* in the second argument.

Remark 3.1.4. For short, we shall often refer to a complex vector space V with a given inner product $(\cdot, \cdot)$ as an *inner product space*.

Example 3.1.5. Denote by $\mathbb{C}^n$ the space of complex n-tuples,

$$\mathbb{C}^n = \big\{(x_1, \ldots, x_n) \,\big|\, x_i \in \mathbb{C}, \; i = 1, \ldots, n\big\} \,.$$

Define $(\cdot, \cdot) : \mathbb{C}^n \times \mathbb{C}^n \to \mathbb{C}$ by

$$(x, y) = \sum_{i=1}^{n} x_i \overline{y_i} \quad \text{for} \quad x, y \in \mathbb{C}^n.$$

Then it is easy to prove that $\mathbb{C}^n$ is a complex vector space with the obvious coordinatewise defined operations and that $(\cdot, \cdot)$ is an inner product in $\mathbb{C}^n$.

Example 3.1.6. In the vector space of continuous functions $C_0(\mathbb{R})$, we can define an inner product by the definition,

$$(f, g) = \int_{\mathbb{R}} f(x)\overline{g(x)} \, dx \quad \text{for} \quad f, g \in C_0(\mathbb{R}) \,.$$

We can extend this inner product to the vector space of quadratic integrable functions $L^2(\mathbb{R})$ by a limit argument.

Assertion 3.1.7. *Suppose that $f, g \in L^2(\mathbb{R})$ are limit functions of the Cauchy sequences $(f_n), (g_n)$ in $(C_0(\mathbb{R}), ||\cdot||_2)$. Then $(\int_{\mathbb{R}} f_n(x)\overline{g_n(x)} \, dx)$ is a Cauchy sequence in $(\mathbb{C}, |\cdot|)$. The sequence has a well-defined limit in $\mathbb{C}$, which for obvious reasons is denoted by $\int_{\mathbb{R}} f(x)\overline{g(x)} \, dx$. By this definition we have*

$$\int_{\mathbb{R}} f(x)\overline{g(x)} \, dx = \lim_{n \to \infty} \int_{\mathbb{R}} f_n(x)\overline{g_n(x)} \, dx \,.$$

Proof. By the Cauchy-Schwarz inequality for continuous functions, Corollary 2.3.7, we get the following estimates,

$$\left| \int_{\mathbb{R}} \Big(f_n(x)\overline{g_n(x)} - f_m(x)\overline{g_m(x)} \Big) dx \right|$$

$$\leq \left| \int_{\mathbb{R}} \Big(f_n(x) - f_m(x) \Big) \overline{g_n(x)} \, dx \right| + \left| \int_{\mathbb{R}} f_m(x) \Big(\overline{g_n(x)} - \overline{g_m(x)} \Big) dx \right|$$

$$\leq ||f_n - f_m||_2 ||g_n||_2 + ||f_m||_2 ||g_n - g_m||_2$$

$$\leq ||f_n - f_m||_2 \, k_1 + k_2 ||g_n - g_m||_2.$$

Note that there exist constants k_1 and k_2 bounding the Cauchy sequences $(||g_n||_2)$ and $(||f_m||_2)$. The estimates prove that $(\int_{\mathbb{R}} f_n(x)\overline{g_n(x)} \, dx)$ is a Cauchy sequence in $(\mathbb{C}, |\cdot|)$. Since $(\mathbb{C}, |\cdot|)$ is complete, the sequence has a well-defined limit, which we denote by $\int_{\mathbb{R}} f(x)\overline{g(x)} \, dx$. $\qquad\square$

We can now immediately extend the inner product in $C_0(\mathbb{R})$ to $L^2(\mathbb{R})$ by the definition

$$(f, g) = \int_{\mathbb{R}} f(x)\overline{g(x)}\, dx \quad \text{for} \quad f, g \in L^2(\mathbb{R}) \ .$$

The norm $||\cdot||_2$ in $L^2(\mathbb{R})$ is given by

$$||f||_2 = \left(\int_{\mathbb{R}} |f(x)|^2 dx \right)^{1/2} = \left(\int_{\mathbb{R}} f(x)\overline{f(x)}\, dx \right)^{1/2} = \sqrt{(f, f)} \ .$$

Unless otherwise specified, we shall always assume that $L^2(\mathbb{R})$ is equipped with the above inner product $(\cdot, \cdot)$ and the related norm $||\cdot||_2$.

Lemma 3.1.8. *Let $(\cdot, \cdot)$ be an inner product in the complex vector space V. Then we have the following inequality*

$$|(x, y)|^2 \leq (x, x)(y, y) \quad \text{for all} \quad x, y \in V \ .$$

Proof. If $(y, y) = 0$, then $y = 0$, and the inequality is trivially true. Suppose therefore that $(y, y) \neq 0$.

For any $\alpha \in \mathbb{C}$ we get

$$0 \leq (x - \alpha y, x - \alpha y) = (x, x) - \alpha(y, x) - \overline{\alpha}(x, y) + \alpha\overline{\alpha}(y, y) \ .$$

Put $\alpha = (x, y)/(y, y)$ and we get

$$0 \leq (x, x) - \frac{(x, y)}{(y, y)}(y, x) - \frac{\overline{(x, y)}}{(y, y)}(x, y) + \frac{(x, y)}{(y, y)}\frac{\overline{(x, y)}}{(y, y)}(y, y)$$

$$= (x, x) - \frac{(x, y)\overline{(x, y)}}{(y, y)} \ .$$

The inequality now follows immediately. $\qquad\qquad\qquad\qquad\qquad\qquad \square$

In Example 3.1.6 it was shown how the 2-norm $||\cdot||_2$ in $L^2(\mathbb{R})$ can be derived from the inner product $(\cdot, \cdot)$ in $L^2(\mathbb{R})$. A similar construction can be applied to derive a norm from an inner product in an arbitrary vector space.

Theorem 3.1.9. *Let V be a complex vector space with an inner product $(\cdot, \cdot)$. Then the inner product induces a norm $||\cdot||$ in V by the definition*

$$||x|| = \sqrt{(x, x)} \quad \text{for} \quad x \in V \ .$$

Proof.   We have to prove that $||\cdot||$ satisfies the three conditions for a norm. NORM 1 follows by IP 3:

$$||x|| = \sqrt{(x, x)} \geq 0, \quad = 0 \Leftrightarrow x = 0 ,$$

for all $x \in V$.

NORM 2 follows by using the linearity properties of the inner product:

$$||\alpha x|| = \sqrt{(\alpha x, \alpha x)} = \sqrt{\alpha \overline{\alpha} (x, x)} = \sqrt{|\alpha|^2 (x, x)} = |\alpha| \, ||x|| ,$$

for all $\alpha \in \mathbb{C}$, and all $x \in V$.

NORM 3, the triangle inequality, follows by a computation using in turn the linearity properties of the inner product, IP 1, and Lemma 3.1.8:

$$\begin{aligned}
||x + y||^2 &= (x + y, x + y) \\
&= ||x||^2 + ||y||^2 + (x, y) + (y, x) \\
&= ||x||^2 + ||y||^2 + (x, y) + \overline{(x, y)} \\
&= ||x||^2 + ||y||^2 + 2\operatorname{Re}[(x, y)] \\
&\leq ||x||^2 + ||y||^2 + 2|(x, y)| \\
&\leq ||x||^2 + ||y||^2 + 2||x|| \, ||y|| \\
&= \left(||x|| + ||y|| \right)^2 ,
\end{aligned}$$

for all $x, y \in V$. $\square$

Remark 3.1.10. The norm $||\cdot||$ defined by the procedure in Theorem 3.1.9 is called the *induced norm* given by the inner product $(\cdot, \cdot)$.

Having proved that the construction in Theorem 3.1.9 actually defines a norm, Lemma 3.1.8 can be reformulated as Cauchy-Schwarz' inequality for a general vector space with an inner product.

Theorem 3.1.11 (Cauchy-Schwarz' inequality). *Let V be a complex vector space with an inner product $(\cdot, \cdot)$ and induced norm $||\cdot||$. Then we have the inequality*

$$|(x, y)| \leq ||x|| \, ||y|| \quad \text{for all} \quad x, y \in V .$$

The inner product is continuous in both variables. This is the message in the following result.

Proposition 3.1.12. *Let $(\cdot, \cdot)$ be an inner product in the complex vector space V and let $||\cdot||$ be the induced norm.*

If the sequences (x_n) and (y_n) in V are convergent to $x \in V$, respectively $y \in V$, then the sequence of inner products $((x_n, y_n))$ converges to (x, y), i.e.

$$\left.\begin{array}{c} x_n \to x \\ y_n \to y \end{array}\right\} \quad \text{for} \quad n \to \infty \implies (x_n, y_n) \to (x, y) \quad \text{for} \quad n \to \infty .$$

Proof. First observe that

$$(x_n, y_n) - (x, y) = (x_n, y_n) - (x, y_n) + (x, y_n) - (x, y)$$
$$= (x_n - x, y_n) + (x, y_n - y)$$
$$= (x_n - x, y_n - y) + (x_n - x, y) + (x, y_n - y) .$$

From this we get by the Cauchy-Schwarz inequality:

$$|(x_n, y_n) - (x, y)| = |(x_n - x, y_n - y) + (x_n - x, y) + (x, y_n - y)|$$
$$\leq |(x_n - x, y_n - y)| + |(x_n - x, y)| + |(x, y_n - y)|$$
$$\leq ||x_n - x|| \, ||y_n - y|| + ||x_n - x|| \, ||y|| + ||x|| \, ||y_n - y|| .$$

The proposition now follows easily. $\qquad\square$

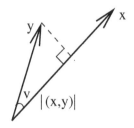

Fig. 3.1   Inner product of vectors in $\mathbb{R}^3$

In 3-dimensional Euclidean space $\mathbb{R}^3$ with the canonical inner product $(x, y) = \sum_{i=1}^{3} x_i y_i$, it is well known that two vectors $x, y \in \mathbb{R}^3$ are orthogonal in the usual geometrical sense of the word, written $x \perp y$, if and only if $(x, y) = 0$. This follows from the alternative formula for the inner product of two vectors $x, y \in \mathbb{R}^3$ forming an angle of v:

$$(x, y) = ||x|| \, ||y|| \cos v .$$

Up to sign, the inner product (x, y) of vectors $x, y \in \mathbb{R}^3$ is the length of the orthogonal projection of y onto x multiplied by the length of x; cf. Figure 3.1.

We use the 3-dimensional Euclidean case to motivate the definition of orthogonality of vectors in arbitrary inner product spaces.

Definition 3.1.13. In a complex vector space V with inner product $(\cdot, \cdot)$, two vectors $x, y \in V$ are said to be *orthogonal*, written $x \perp y$, if $(x, y) = 0$.

For orthogonal vectors we have an abstract version of Pythagoras' theorem.

Theorem 3.1.14 (Pythagoras' theorem). *Suppose that x and y is an arbitrary pair of orthogonal vectors in the inner product space V, i.e. suppose $x \perp y$. Then we have Pythagoras' formula*

$$||x+y||^2 = ||x||^2 + ||y||^2 .$$

Proof.   The proof follows by the easy computation

$$||x+y||^2 = (x+y, x+y) = ||x||^2 + ||y||^2 + (x, y) + (y, x) = ||x||^2 + ||y||^2 ,$$

since $(x, y) = (y, x) = 0$. $\square$

3.2   Hilbert spaces

Let V be a complex vector space with an inner product $(\cdot, \cdot)$ and the induced norm $||\cdot||$. If V is a Banach space, in other words a complete normed vector space, with respect to the induced norm $||\cdot||$, it is called a Hilbert space.

Definition 3.2.1. A *Hilbert space* is an inner product space V which is a Banach space with respect to the induced norm.

Example 3.2.2. Here are some main examples of Hilbert spaces.
    1. $\mathbb{C}^n$ with the standard inner product is a Hilbert space. It boils down to the fact that $\mathbb{R}^{2n}$ with Euclidean norm is a complete space.
    2. $L^2(\mathbb{R})$ with the inner product

$$(f, g) = \int_{\mathbb{R}} f(x)\overline{g(x)}\, dx ,$$

inducing the 2-norm $||\cdot||_2$, is by construction a complete space and hence a Hilbert space.
    3. l^2, the space of square summable complex sequences (x_n), i.e. sequences for which $\sum_{n=1}^{\infty} |x_n|^2 < \infty$, is a Hilbert space with the inner product

$$((x_n), (y_n)) = \sum_{n=1}^{\infty} x_n \overline{y_n} .$$

Note that the series $\sum_{n=1}^{\infty} x_n \overline{y_n}$ is convergent, so that the inner product is defined. This is an immediate consequence of the Cauchy-Schwarz inequality:

$$\sum_{n=1}^{\infty} |x_n y_n| \leq ||(x_n)||_2 ||(y_n)||_2 \, .$$

The properties required for an inner product are easily established. And so is completeness of the space.

3.3 Basis in a normed vector space and separability

Several constructions in functional analysis involve objects defined as sums of infinite series in normed vector spaces. Fortunately the basic theory of such infinite series is completely parallel to the theory of series with complex numbers as terms.

3.3.1 *Infinite series in normed vector spaces*

Let V be a normed vector space with norm $||\cdot||$.

By a *series* $\sum_{n=1}^{\infty} x_n$ in V, we understand a sequence $x_1, x_2, \ldots, x_n, \ldots$ of elements of V which we, to begin with, just formally sum. To the series there corresponds a sequence of *partial sums* $s_1, s_2, \ldots, s_n, \ldots$ defined by

$$s_n = \sum_{k=1}^{n} x_k = x_1 + \cdots + x_n.$$

We say that the series is *convergent* with *sum* $s \in V$ if the sequence of partial sums (s_n) is convergent with limit point s. In case of convergence we write

$$s = \sum_{n=1}^{\infty} x_n.$$

The nth term x_n in a series $\sum_{n=1}^{\infty} x_n$ can be written in the form $x_n = s_n - s_{n-1}$, which gives the following necessary condition for convergence of a series.

Theorem 3.3.1. *In a convergent series $\sum_{n=1}^{\infty} x_n$ in the normed vector space V, the nth term x_n converges to $0 \in V$ for n going to ∞.*

Since a Banach space considered as a metric space is complete, we also have the following important result.

Theorem 3.3.2 (The Cauchy condition for series). *A series $\sum_{n=1}^{\infty} x_n$ in a Banach space V is convergent if and only if*

$$\forall \varepsilon > 0 \; \exists n_0 \in \mathbb{N} \; \forall n, k \in \mathbb{N} \; : \quad n \geq n_0 \quad \Rightarrow \|s_{n+k} - s_n\| < \varepsilon,$$

or equivalently,

$$\forall \varepsilon > 0 \; \exists n_0 \in \mathbb{N} \; \forall n, k \in \mathbb{N} \; : \quad n \geq n_0 \quad \Rightarrow \left\| \sum_{i=n+1}^{n+k} x_i \right\| < \varepsilon.$$

From the Cauchy condition we deduce a very useful test for convergence.

Theorem 3.3.3 (The comparison test). *Let $\sum_{n=1}^{\infty} x_n$ be a series in a Banach space V. If there exists a convergent series of non-negative real numbers $\sum_{n=1}^{\infty} a_n$ such that $\|x_n\| \leq a_n$ for $n \geq n_0$ from a certain step n_0, then the series $\sum_{n=1}^{\infty} x_n$ is convergent.*

Proof. For $n \geq n_0$, the triangle inequality gives the inequality

$$\left\| \sum_{i=n+1}^{n+k} x_i \right\| \leq \sum_{i=n+1}^{n+k} a_i.$$

Since the Cauchy condition is satisfied in both V and $\mathbb{R}$, the result follows immediately. $\square$

A series $\sum_{n=1}^{\infty} x_n$ is said to be *divergent*, if it is not convergent.

3.3.2   Separability of a normed vector space

Let V be a normed vector space with norm $\|\cdot\|$.

If there exists a finite set $\{x_1, \ldots, x_n\}$ of elements in V with the property that every element $x \in V$ can be written in a unique manner as a linear combination,

$$x = \alpha_1 x_1 + \cdots + \alpha_n x_n \; , \; \alpha_i \in \mathbb{C} \; ,$$

then it is well known from basic linear algebra that the number n is uniquely determined. We say that V is *finite dimensional* of *dimension n* and that $\{x_1, \ldots, x_n\}$ is a *basis* for V.

If there exists a countable set $\{x_1, \ldots, x_n, \ldots\}$ of elements in V, in other words a sequence (x_n) in V, then we say that the sequence (x_n) is a *Schauder basis* in V, if every element $x \in V$ admits a unique decomposition as the sum of a convergent series

$$x = \sum_{n=1}^{\infty} \alpha_n x_n , \quad \alpha_n \in \mathbb{C}, \; \forall n \in \mathbb{N} .$$

If (x_n) is a Schauder basis in V it follows in particular that

$$x = \sum_{n=1}^{\infty} \alpha_n x_n = 0 \quad \Longleftrightarrow \quad \alpha_n = 0, \; \forall n \in \mathbb{N} .$$

This implies that every finite subset of elements in the basis (x_n) is a linearly independent set of elements in V. Hence a normed vector space V with a Schauder basis is necessarily of infinite dimension since it contains subspaces of an arbitrary finite dimension.

Definition 3.3.4. We call a normed vector space V *separable by a basis* if V admits a finite basis (finite dimension) or a Schauder basis (infinite dimension).

For topological spaces, there is a more general notion of separability.

Definition 3.3.5. A topological space S is called *separable* if it admits a countable dense subset W.

Definition 3.3.5 applies in particular to metric spaces and hence also to normed vector spaces. For normed vector spaces we then have two notions of separability. As the following theorem shows, separability by a basis is a stronger notion than just separability.

Theorem 3.3.6. *If a normed vector space V is separable by a basis, then it is separable.*

Proof. If V is a finite dimensional normed vector space, we consider the set of points W in V defined by all linear combinations of a finite set of basis vectors for V, using only such complex numbers as coefficients for which both the real part and the imaginary part are rational numbers. This set of points W is countable, since the set of rational numbers is countable. The set W is furthermore dense in V, since the set of rational numbers is dense in the set of real numbers.

If V admits a Schauder basis, we consider by analogy the set of points W in V defined by all finite linear combinations of the set of vectors in a Schauder basis for V, using only such complex numbers as coefficients for which both the real part and the imaginary part are rational numbers. Also in this case, the set of points W is countable and dense in V. This completes the proof. $\square$

Remark 3.3.7. In the case of Hilbert spaces, separability by a basis coincides with the general notion of separability; cf. Theorem 3.3.6 and Theorem 3.4.9. For Banach spaces, the two notions of separability do not agree. After it had been an open problem for 40 years, the Swedish mathematician Per Enflo in 1973 gave examples of Banach spaces that contain a countable dense subset of points but do not admit a Schauder basis. In this context, further remarkable new discoveries about Banach spaces were made towards the end of the 20th century by the British mathematician William Timothy Gowers, who was awarded a Fields medal for his work in 1998.

All normed vector spaces considered in this book will in practice be finite dimensional, or, at least separable, infinite dimensional. This will be sufficient to cover most applications.

Example 3.3.8. Let l^2 be the space of square summable sequences. Then l^2 is an infinite dimensional, separable Hilbert space, and the sequence (e_n) in l^2, where $e_n = (\ldots, 1, \ldots)$ is the sequence in $\mathbb{C}$ with 1 at the nth place and 0 otherwise, is a Schauder basis for l^2.

## 3.4	Basis in a separable Hilbert space

The basis (e_n) in l^2 in Example 3.3.8 is of a particular convenient type. It is an orthonormal basis.

Definition 3.4.1. A set of vectors $\{x_k\}$ in an inner product space is called an *orthogonal set* if $(x_i, x_j) = 0$ for $i \neq j$, and $x_j \neq 0$ for all j. If furthermore $(x_j, x_j) = 1$ for all j, in other words, if all the vectors are unit vectors, then the set is called an *orthonormal set*. A sequence (x_n), which as a set $\{x_n\}$ is orthonormal, is called an *orthonormal sequence*.

If an orthonormal set of vectors in an inner product space is a basis for the vector space, then we call the set an *orthonormal basis* - a terminology that applies both in the finite dimensional case and in the separable, infinite dimensional case.

Proposition 3.4.2. *An arbitrary finite orthogonal set in an inner product space is linearly independent.*

Proof. Suppose that $\{e_1, \ldots, e_n\}$ is an orthogonal set in an inner product space and that we are given a vanishing linear combination

$$\alpha_1 e_1 + \cdots + \alpha_n e_n = 0 .$$

Then

$$\alpha_j(e_j, e_j) = (\alpha_1 e_1 + \cdots + \alpha_n e_n, e_j) = 0 ,$$

and hence $\alpha_j = 0$, for each $j = 1, \ldots, n$. This proves that the set $\{e_1, \ldots, e_n\}$ is linearly independent. □

Orthonormal sets are very convenient to use as a basis and hence it is fortunate that we have an algorithm by which to orthonormalize a given set of linearly independent vectors.

Theorem 3.4.3 (The Gram-Schmidt procedure). *Let $\{y_k\}$ be a (finite, or countable) set of linearly independent vectors in a Hilbert space H. Then there is an orthonormal set $\{x_k\}$ in H, such that for any $n \in \mathbb{N}$ we have*

$$\operatorname{span}\{x_k\}_{k=1}^n = \operatorname{span}\{y_k\}_{k=1}^n .$$

Proof. Define recursively

$$x_1 = \frac{y_1}{\|y_1\|} , \quad x_2 = \frac{y_2 - (y_2, x_1)x_1}{\|y_2 - (y_2, x_1)x_1\|} ,$$

and if we assume that $x_1, x_2, \ldots, x_j$ are defined,

$$x_{j+1} = \frac{y_{j+1} - \sum_{k=1}^j (y_{j+1}, x_k)x_k}{\|y_{j+1} - \sum_{k=1}^j (y_{j+1}, x_k)x_k\|} .$$

Then the set (x_k) is orthonormal by construction and it satisfies the requirement about span. □

We shall now investigate the possibilities for expansion of the vectors in an infinite dimensional, separable Hilbert space into infinite series using a sequence of basis vectors. As a culmination we shall prove that by choosing an orthonormal basis in an infinite dimensional, separable Hilbert space H, we can map H onto l^2 by an isomorphism preserving the inner products in the two spaces (an isometry), whereby H is in fact identified with l^2. The expansion of a vector in H using an orthonormal basis is a generalization of the classical *Fourier expansion* of L^2-functions.

Proposition 3.4.4. *Let (x_n) be an orthonormal sequence in the Hilbert space H, and let (α_n) be a sequence of real, or complex, numbers. Then the series $\sum_{n=1}^\infty \alpha_n x_n$ is convergent in H if and only if $\sum_{n=1}^\infty |\alpha_n|^2$ is convergent in $\mathbb{R}$.*

In case of convergence, it holds that

$$\left\|\sum_{n=1}^{\infty} \alpha_n x_n\right\|^2 = \sum_{n=1}^{\infty} |\alpha_n|^2 .$$

Proof. Consider the sequence of partial sums (s_n),

$$s_n = \sum_{i=1}^{n} \alpha_i x_i .$$

Since H is a Hilbert space, the series $\sum_{n=1}^{\infty} \alpha_n x_n$ is convergent in H if and only if (s_n) is a Cauchy sequence in H. Now

$$\|s_{n+k} - s_n\|^2 = \left(\sum_{i=n+1}^{n+k} \alpha_i x_i, \sum_{i=n+1}^{n+k} \alpha_i x_i\right) = \sum_{i=n+1}^{n+k} |\alpha_i|^2 .$$

Since the Cauchy condition is satisfied for series of real numbers it follows that (s_n) is a Cauchy sequence in H if and only if the series $\sum_{n=1}^{\infty} |\alpha_n|^2$ is convergent in $\mathbb{R}$.

Moreover, in case of convergence, we get using Proposition 3.1.12,

$$\left\|\sum_{n=1}^{\infty} \alpha_n x_n\right\|^2 = \left\|\lim_{n\to\infty} s_n\right\|^2 = \lim_{n\to\infty} \|s_n\|^2 = \sum_{n=1}^{\infty} |\alpha_n|^2 .$$

$\square$

The following important result is abstracted from work of the German mathematician Friedrich Wilhelm Bessel (1784–1846).

Theorem 3 4 5 (Bessel's equation and inequality). *Let (x_n) be an arbitrary orthonormal sequence in the Hilbert space H. Then for all $x \subset H$ and $n \in \mathbb{N}$ we have the equation,*

$$\left\|x - \sum_{k=1}^{n} (x, x_k) x_k\right\|^2 = \|x\|^2 - \sum_{k=1}^{n} |(x, x_k)|^2 .$$

From the equation we get the inequality $\sum_{k=1}^{n} |(x, x_k)|^2 \le \|x\|^2$, and

$$\sum_{k=1}^{\infty} |(x, x_k)|^2 \le \|x\|^2 .$$

Conclusion: The series $\sum_{k=1}^{\infty} (x, x_k) x_k$ is convergent for all $x \in H$.

Proof. Put $x_0 = \sum_{k=1}^{n} (x, x_k) x_k$ and $y_0 = x - x_0$.

A small computation shows that $(x_0, y_0) = 0$, since

$$(x_k, y_0) = (x_k, x) - (x_k, x_0)$$
$$= (x_k, x) - \overline{(x, x_k)}$$
$$= (x_k, x) - (x_k, x) = 0,$$

for all $k = 1, \ldots, n$.

By Pythagoras' theorem, Theorem 3.1.14, we then get

$$||x||^2 = ||x_0 + y_0||^2$$
$$= ||x_0||^2 + ||y_0||^2$$
$$= \sum_{k=1}^{n} |(x, x_k)|^2 + ||x - \sum_{k=1}^{n}(x, x_k)||^2,$$

which proves Bessel's equation.

The remaining part of the theorem now follows easily. For the final conclusion we use Proposition 3.4.4. □

Using the same idea of proof we get the following useful result.

Theorem 3.4.6 (Best Approximation Theorem). *Let (x_n) be an arbitrary orthonormal sequence in the Hilbert space H, and let $\alpha_1, \ldots, \alpha_n$ be a set of real, or complex, numbers. Then*

$$||x - \sum_{k=1}^{n} \alpha_k x_k|| \geq ||x - \sum_{k=1}^{n}(x, x_k)x_k||.$$

Proof. As before, put $x_0 = \sum_{k=1}^{n}(x, x_k)x_k$ and $y_0 = x - x_0$. Furthermore, put $z_0 = \sum_{k=1}^{n} \alpha_k x_k$.

Proceeding as before we see that also $(x_0 - z_0, y_0) = 0$.

Then we get by Pythagoras' theorem,

$$||x - z_0||^2 = ||x - x_0 + x_0 - z_0||^2$$
$$= ||y_0 + x_0 - z_0||^2$$
$$= ||y_0||^2 + ||x_0 - z_0||^2 \geq ||y_0||^2,$$

which proves the inequality. □

An important consequence of the Best Approximation Theorem is that an orthonormal basis for a dense subspace of a Hilbert space is actually an orthonomal basis in the full Hilbert space. This is a very useful result for the construction of specific orthonormal basis in separable Hilbert spaces. The precise result is as follows.

Theorem 3.4.7. *Let V be a dense subspace of the Hilbert space H, and assume that $\{e_n\}$ is an orthonormal basis for V (finite, or countable). Then $\{e_n\}$ is also an orthonormal basis for H.*

Proof.   The finite dimensional case is easy. A linear subspace V in H of finite dimension is a complete space, thus a closed set in H and hence $V = \overline{V} = H$.

Then to the infinite dimensional case. Since (e_n) is now a Schauder basis for V, any vector $v \in V$ admits a unique expansion as an infinite series $v = \sum_{n=1}^{\infty} \alpha_n e_n$. In fact, $v = \sum_{n=1}^{\infty} (v, e_n) e_n$. This follows by the short computation

$$(v, e_i) = \lim_{n \to \infty} \left(\sum_{k=1}^{n} \alpha_k e_k, e_i \right) = \alpha_i \ ,$$

valid for all $i \in \mathbb{N}$.

Let $x \in H$ be an arbitrary vector in H. We have to prove that x admits a similar expansion as an infinite series in terms of (e_n).

Assertion 3.4.8. *An arbitrary vector $x \in H$ admits the expansion*

$$x = \sum_{n=1}^{\infty} (x, e_n) e_n \ .$$

To prove the assertion, let an arbitrary $\varepsilon > 0$ be given. Since V is dense in H, we can (and do) choose $v \in V$, such that $\|x - v\| < \varepsilon/2$.

Now write $v = \sum_{n=1}^{\infty} (v, e_n) e_n$, and choose $n_0 \in \mathbb{N}$ such that

$$n \geq n_0 \Rightarrow \left\| v - \sum_{k=1}^{n} (v, e_k) e_k \right\| < \varepsilon/2 \ .$$

By the Best Approximation Theorem we then get for all $n \geq n_0$,

$$\left\| x - \sum_{k=1}^{n} (x, e_k) e_k \right\| \leq \left\| x - \sum_{k=1}^{n} (v, e_k) e_k \right\|$$

$$\leq \| x - v \| + \left\| v - \sum_{k=1}^{n} (v, e_k) e_k \right\|$$

$$< \varepsilon/2 + \varepsilon/2 = \varepsilon \ .$$

This proves that

$$x = \lim_{n \to \infty} \sum_{k=1}^{n} (x, e_k)e_k = \sum_{n=1}^{\infty} (x, e_n)e_n \ ,$$

which is exactly the statement in the assertion.

This completes the proof of the theorem. $\square$

We can now prove the following basic result for separable Hilbert spaces.

Theorem 3.4.9. *Every separable Hilbert space H has an orthonormal basis.*

Proof. The finite dimensional case is immediate. Just apply the Gram-Schmidt procedure, Theorem 3.4.3, to an arbitrary finite set of basis vectors for H.

Then to the infinite dimensional, separable case.

By the definition of separability, H contains a countable, dense subset W of vectors in H. Consider the linear subspace U in H consisting of all finite linear combinations of vectors in W - the *linear span* of W. Clearly, U is a dense subspace in H. By the construction of U we can eliminate vectors from the countable set W one after the other to get a linearly independent set $\{x_n\}$ (finite, or countable) of vectors in U that spans U.

If the set $\{x_n\}$ is finite we get back to the finite dimensional case.

If the set $\{x_n\}$ is infinite, we use the Gram-Schmidt procedure, to turn the sequence (x_n) into an orthonormal sequence (y_n) with the property that for all $n \in \mathbb{N}$,

$$\mathrm{span}\{x_k\}_{k=1}^{n} = \mathrm{span}\{y_k\}_{k=1}^{n} \ .$$

Now use (y_n) as a Schauder basis to generate a linear subspace V in H. Then V is a dense linear subspace of H, since U is a dense linear subspace of V. The latter follows since any vector $y \in V$ can be expanded into a series $y = \sum_{k=1}^{\infty} \alpha_k y_k$, showing that $y = \lim_{n \to \infty} \sum_{k=1}^{n} \alpha_k y_k$, and hence that y is the limit of a sequence of vectors in U.

By construction, (y_n) is an orthonormal basis for V and hence by Theorem 3.4.7 also for H. $\square$

We have the following convenient characterization of an orthonormal basis.

Theorem 3.4.10. *For an orthonormal sequence (x_n) in a Hilbert space H, the following conditions are equivalent:*

(1) *(x_n) is an orthonormal basis in H.*

(2) *For all $x, y \in H$, the inner product (x, y) satisfies*

$$(x, y) = \sum_{n=1}^{\infty} (x, x_n)(x_n, y) .$$

(3) *For all $x \in H$, the norm $\|x\|$ satisfies* Parseval's equation,

$$\|x\|^2 = \sum_{n=1}^{\infty} |(x, x_n)|^2 .$$

(4) *The sequence (x_n) is a total set, i.e.*

$$(x, x_n) = 0, \ \forall n \in \mathbb{N} \quad \Rightarrow \quad x = 0 .$$

Proof. We prove the theorem by proving the following closed cycle of implications: (1) $\Rightarrow$ (2) $\Rightarrow$ (3) $\Rightarrow$ (4) $\Rightarrow$ (1).

(1) $\Rightarrow$ (2). If (x_n) is an orthonormal basis in H, arbitrary vectors $x, y \in H$
admit expansions

$$x = \sum_{n=1}^{\infty} (x, x_n)x_n \quad \text{and} \quad y = \sum_{n=1}^{\infty} (y, x_n)x_n .$$

By continuity of the inner product, we then get

$$(x, y) = \lim_{n \to \infty} \left(\sum_{k=1}^{n} (x, x_k)x_k, \ \sum_{k=1}^{n} (y, x_k)x_k \right)$$

$$= \lim_{n \to \infty} \sum_{k-1}^{n} (x, x_k)\overline{(y, x_k)}$$

$$= \sum_{n=1}^{\infty} (x, x_n)(x_n, y) ,$$

which proves (2).

(2) $\Rightarrow$ (3). Put $y = x$ in (2).

(3) $\Rightarrow$ (4). If $(x, x_n) = 0$, all $n \in \mathbb{N}$, we get from (3) that $\|x\| = 0$ and
hence that $x = 0$.

(4) $\Rightarrow$ (1). Let $x \in H$ be an arbitrary vector in H. By Bessel's inequality,
Theorem 3.4.5, it follows that the series $\sum_{n=1}^{\infty} (x, x_n)x_n$ is convergent in H
and thus defines the vector

$$y = \sum_{n=1}^{\infty} (x, x_n)x_n \quad \text{in} \quad H .$$

Since

$$(x - y, x_n) = (x, x_n) - (y, x_n) = (x, x_n) - (x, x_n) = 0$$

for all $n \in \mathbb{N}$, it follows by (4) that $x - y = 0$, i.e. $x = y$. This proves that $x = \sum_{n=1}^{\infty}(x, x_n)x_n$. Since $x \in H$ was arbitrarily chosen this proves that (x_n) is an orthonormal basis in H.

This completes the proof of the theorem. □

We are now ready for the proof of a main result on the structure of infinite dimensional, separable Hilbert spaces: The space of square summable sequences l^2 is the canonical model space!

Theorem 3.4.11 (Riesz-Fischer). *Every separable Hilbert space H of infinite dimension is isometric isomorphic to the Hilbert space l^2 of square summable sequences.*

Proof. Let (e_n) be an orthonormal basis in H. Then every vector $x \in H$ admits the unique expansion $x = \sum_{k=1}^{\infty}(x, e_k)e_k$. By Parseval's equation, Theorem 3.4.10, the sequence $((x, e_n))$ is square summable, in other words an element of l^2. We can therefore define a map

$$T : H \to l^2 \quad \text{by} \quad T(x) = ((x, e_n)) \ .$$

Clearly T is linear.

T is injective, since $T(x) = 0$ implies $(x, e_n) = 0$ for all $n \in \mathbb{N}$, and hence $x = 0$ by Theorem 3.4.10.

T is surjective, since for any vector $(\alpha_n) \in l^2$, the series $\sum_{k=1}^{\infty}\alpha_k e_k$ is convergent (Proposition 3.4.4), thus defining a vector $x = \sum_{k=1}^{\infty}\alpha_k e_k$ in H for which

$$T(x) = ((x, e_n)) = (\alpha_n) \ .$$

From Parseval's equation we have

$$\|T(x)\|^2 = \sum_{n=1}^{\infty}|(x, e_n)|^2 = \|x\|^2 \ ,$$

such that $\|T(x)\| = \|x\|$ for all $x \in H$. This proves that T is an isometry. It also proves that T is a bounded linear isomorphism.

Also the inner product is preserved. This follows by the computation

$$(T(x), T(y)) = \Big(\big((x, e_n) \big), \big((y, e_n) \big) \Big) = \sum_{n=1}^{\infty} (x, e_n) \overline{(y, e_n)}$$

$$= \sum_{n=1}^{\infty} (x, e_n)(e_n, y) = (x, y) \,,$$

using property (2) in Theorem 3.4.10. □

3.5　Orthogonal projection and complement

Let (e_n) be an orthonormal basis in a Hilbert space H and define for each $n \in \mathbb{N}$ the subspace

$$E_n = \mathrm{span}\{e_1, \ldots, e_n\} \,.$$

For an arbitrary vector $x \in H$, consider the vector $s_n = \sum_{k=1}^{n} (x, e_k) e_k \in E_n$. The vector s_n is the *orthogonal projection* of x onto E_n, since $(x - s_n) \perp e_k$ for each $k = 1, \ldots, n$. The vector s_n is also the vector in E_n closest to x. As the proof of the Best Approximation Theorem shows, the vector $y = s_n$ is, in fact, the unique solution to the problem of minimizing the norm $\|x - y\|$ for $y \in E_n$. As we shall now prove, there is a corresponding result when E_n is substituted by a closed convex subset K of H, in particular an arbitrary closed linear subspace of H. To establish this, we need the so-called Parallelogram law.

Proposition 3.5.1 (The Parallelogram law). *In a normed vector space V, where the norm $\|\cdot\|$ is induced by an inner product $(\cdot, \cdot)$, it holds that*

$$\|x + y\|^2 + \|x - y\|^2 = 2(\|x\|^2 + \|y\|^2) \,,$$

for all $x, y \in V$.

Proof.　The formula follows immediately by adding the results of the following computations

$$\|x + y\|^2 = (x + y, x + y) = \|x\|^2 + (x, y) + (y, x) + \|y\|^2,$$
$$\|x - y\|^2 = (x - y, x - y) = \|x\|^2 - (x, y) - (y, x) + \|y\|^2.$$

□

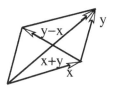

Fig. 3.2   Illustration of Parallelogram law in the Euclidean plane

Theorem 3.5.2. *Let K be a closed convex subset of the Hilbert space H. For every vector $x_0 \in H$, there is a uniquely determined vector $y_0 \in K$, such that*

$$\| x_0 - y_0 \| \leq \| x_0 - y \| ,$$

for all $y \in K$.

Proof.   Put $\delta = \inf\{ \| x_0 - y \| \mid y \in K \}$, and choose a sequence (y_k) in K such that $\| x_0 - y_k \| \to \delta$ for $k \to \infty$.

Assertion 3.5.3. *The sequence (y_k) is a Cauchy sequence.*

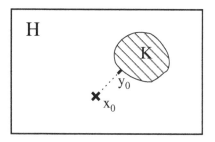

Fig. 3.3   Orthogonal projection onto convex set

Proof.   By the parallelogram law we have

$$\| y_n - y_m \|^2 = \| (y_n - x_0) + (x_0 - y_m) \|^2$$

$$= 2\| y_n - x_0 \|^2 + 2\| x_0 - y_m \|^2 - \| (y_n - x_0) - (x_0 - y_m) \|^2$$

$$= 2\| y_n - x_0 \|^2 + 2\| x_0 - y_m \|^2 - 4\| \frac{1}{2}(y_n + y_m) - x_0 \|^2$$

$$\leq 2\| y_n - x_0 \|^2 + 2\| x_0 - y_m \|^2 - 4\delta^2 ,$$

since $\frac{1}{2}(y_n + y_m) \in K$ by the convexity assumption on K.

From this inequality follows that

$$||y_n - y_m||^2 \to 2\delta^2 + 2\delta^2 - 4\delta^2 = 0 \quad \text{for} \quad n, m \to \infty .$$

This proves that (y_k) is a Cauchy sequence as asserted.   □

By completeness of H, the Cauchy sequence (y_k) has a limit point $y_0 \in H$, and since K is closed we have that $y_0 \in K$. In other words, $y_k \to y_0 \in K$ for $k \to \infty$. Then

$$||x_0 - y_0|| = \lim_{k \to \infty} ||x_0 - y_k|| = \delta .$$

Now assume that $\delta = ||x_0 - z_0||$ also for the vector $z_0 \in K$. Then we have the computation

$$\begin{aligned}
||y_0 - z_0||^2 &= ||(y_0 - x_0) + (x_0 - z_0)||^2 \\
&= 2||y_0 - x_0||^2 + 2||x_0 - z_0||^2 - ||(y_0 - x_0) - (x_0 - z_0)||^2 \\
&= 2||y_0 - x_0||^2 + 2||x_0 - z_0||^2 - 4||\frac{1}{2}(y_0 + z_0) - x_0||^2 \\
&\leq 2\delta^2 + 2\delta^2 - 4\delta^2 = 0 ,
\end{aligned}$$

which implies that $y_0 = z_0$.

This completes the proof of the theorem.   □

In case K is a closed linear subspace in H, the minimizer $y_0 \in K$ realizing the minimal distance $||x_0 - y||$ from x_0 to a point $y \in K$, is just the orthogonal projection of x_0 onto K. For that reason, we also call the minimizer y_0 in Theorem 3.5.2 for the *orthogonal projection* of x_0 onto K in case K is just a closed convex subset of H.

Definition 3.5.4. Let M and N be nonempty subsets of the Hilbert space H. We say that M and N are *orthogonal*, and write $M \perp N$, if $(x, y) = 0$ for all $x \in M$ and $y \in N$.

For a non-empty subset M of H, we define the *orthogonal complement* $M^\perp$ to M by

$$M^\perp = \{y \in H \,|\, (x, y) = 0 \text{ for all } x \in M\} .$$

Lemma 3.5.5. *The orthogonal complement $M^\perp$ of a subset M in a Hilbert space H is a closed linear subspace of H.*

Proof.   That $M^\perp$ is a linear subspace of H follows by observing that for arbitrary vectors $y_1, y_2 \in M^\perp$, and arbitrary scalars $\alpha_1, \alpha_2 \in \mathbb{C}$, the linear combination $\alpha_1 y_1 + \alpha_2 y_2$ belongs to $M^\perp$, since

$$(x, \alpha_1 y_1 + \alpha_2 y_2) = \overline{\alpha_1} (x, y_1) + \overline{\alpha_2} (x, y_2) = 0 ,$$

for all $x \in M$.

That $M^\perp$ is a closed set follows by observing that if $y_n \to y$ for $n \to \infty$ and $y_n \in M^\perp$ for all $n \in \mathbb{N}$, then $y \in M^\perp$, since

$$(x, y) = \lim_{n \to \infty} (x, y_n) = 0 ,$$

for all $x \in M$. □

Example 3.5.6. Let (e_n) be an orthonormal basis for the separable Hilbert space H, and let $I \subseteq \mathbb{N}$ be an arbitrary subset of indexes. Then

$$M = \overline{\text{span}\{e_k \,|\, k \in I\}}$$

is a closed linear subspace of H. It is easy to prove that

$$M^\perp = \{y \in H \,|\, (e_k, y) = 0 \text{ for all } k \in I\} .$$

Definition 3.5.7. Let M and N be closed linear subspaces of the Hilbert space H, with $M \perp N$. Then we define the *orthogonal sum* $M \oplus N$ of M and N by

$$M \oplus N = \{z \in H \,|\, z = x + y, \ x \in M, y \in N\} .$$

Lemma 3.5.8. *For the orthogonal sum $M \oplus N$ of two closed, orthogonal linear subspaces M and N of the Hilbert space H it holds that*
  (a) *The representation $z = x + y$ in $M \oplus N$ is unique.*
  (b) *$M \oplus N$ is a closed linear subspace of H.*

Proof. To prove (a), suppose that $z = x_1 + y_1 = x_2 + y_2$, where $x_1, x_2 \in M$, $y_1, y_2 \in N$. Then $x_1 - x_2 = y_2 - y_1$ is a vector that belongs to both M and N. Since $M \perp N$, it now follows that $(x_1 - x_2, x_1 - x_2) = (y_1 - y_2, y_1 - y_2) = 0$ and therefore that $x_1 = x_2$ and $y_1 = y_2$.

Next we prove (b). Clearly $M \oplus N$ is a linear subspace of H. To prove that $M \oplus N$ is a closed subspace, let (z_n) be a convergent sequence of vectors $z_n = x_n + y_n \in M \oplus N$ with limit point $z_0 \in H$. We have to prove that $z_0 \in M \oplus N$. By Pythagoras' theorem

$$\| z_n - z_m \|^2 = \| (x_n - x_m) + (y_n - y_m) \|^2 = \| x_n - x_m \|^2 + \| y_n - y_m \|^2 .$$

From this follows easily that the sequences (x_n) in M and (y_n) in N are both Cauchy sequences. Hence they are convergent in the Hilbert space H, say

$x_n \to x_0$ for $n \to \infty$ and $y_n \to y_0$ for $n \to \infty$. Since both M and N are closed subspaces in H it follows that $x_0 \in M$ and $y_0 \in N$. Furthermore,

$$z_0 = \lim_{n \to \infty} z_n = \lim_{n \to \infty} (x_n + y_n) = \lim_{n \to \infty} x_n + \lim_{n \to \infty} y_n = x_0 + y_0 \ .$$

This completes the proof of the lemma. $\square$

In finite dimensions we have the well-known splitting $\mathbb{R}^n = \mathbb{R}^{n-k} \oplus \mathbb{R}^k$, for $0 < k < n$, obtained by identifying $\mathbb{R}^{n-k}$ with the subspace of $\mathbb{R}^n$, where the last k coordinates are zero, and $\mathbb{R}^k$ with the subspace of $\mathbb{R}^n$, where the first $n - k$ coordinates are zero. In the Hilbert space setting we have a similar result, known as *the projection theorem*.

Theorem 3.5.9 (The Projection Theorem). *If M is a closed linear subspace of the Hilbert space H, then $H = M \oplus M^{\perp}$.*

Proof.   Let $z \in H$. According to the general orthogonal projection result for closed convex subsets of H, Theorem 3.5.2, there exists a unique $x \in M$, such that $\|z - x\| \leq \|z - \tilde{x}\|$ for all $\tilde{x} \in M$. Put $y = z - x$. We now only have to prove that $y \in M^{\perp}$.

If $M = \{0\}$ this is obvious. If $M \neq \{0\}$, we consider the following computation valid for all $\lambda \in \mathbb{C}$ and $x' \in M$,

$$\begin{aligned}
\|y\|^2 &= \|z - x\|^2 \\
&\leq \|z - x - \lambda x'\|^2 \\
&= \|y - \lambda x'\|^2 = (y - \lambda x', y - \lambda x') \\
&= \|y\|^2 - \overline{\lambda}(y, x') - \lambda(x', y) + |\lambda|^2 \|x'\|^2 \ .
\end{aligned}$$

By taking $\lambda = (y, x')$ and choosing x' with $\|x'\| = 1$, we get

$$\|y\|^2 \leq \|y\|^2 - |\lambda|^2 - |\lambda|^2 + |\lambda|^2 = \|y\|^2 - |\lambda|^2 \ .$$

This shows that necessarily $(y, x') = \lambda = 0$ for all $x' \in M$ with $\|x'\| = 1$, and hence that $(y, x') = 0$ for all $x' \in M$. In other words, $y \in M^{\perp}$, and the proof is complete. $\square$

We can now prove one of the most important results in Hilbert space theory, namely that for an arbitrary Hilbert space H, the dual space H^* of continuous linear functionals on H can be identified with H via the inner product $(\cdot, \cdot)$ in H. The theorem is due to the Hungarian mathematician Frigyes Riesz (1880–1956), one of the pioneers in functional analysis.

Theorem 3.5.10 (Riesz' Representation Theorem). *Let $\phi : H \to \mathbb{C}$ be a continuous linear functional on a Hilbert space H. Then there is a unique vector $z \in H$ such that $\phi(x) = (x, z)$ for all $x \in H$.*

Proof. Let $N = \ker\phi = \{x \in H \mid \phi(x) = 0\}$ be the *kernel* of ϕ. Then N is a linear subspace of H, and since ϕ is continuous, N is closed in H.

If $N = H$, the vector $z = 0$, and only this vector, will do.

If $N \neq H$, we can write $H = N \oplus N^\perp$ according to Theorem 3.5.9. Choose a vector $y_0 \in N^\perp$ with $\|y_0\| = 1$. Then surely $\phi(y_0) \neq 0$. For any vector $x \in H$, we can then write

$$\phi(x) = \frac{\phi(x)}{\phi(y_0)} \phi(y_0) = \phi\Big(\frac{\phi(x)}{\phi(y_0)} y_0 \Big) .$$

This implies that

$$\phi\Big(x - \frac{\phi(x)}{\phi(y_0)} y_0 \Big) = 0 ,$$

which shows that

$$x - \frac{\phi(x)}{\phi(y_0)} y_0 \in N .$$

Hence

$$0 = (x - \frac{\phi(x)}{\phi(y_0)} y_0 , y_0) = (x , y_0) - \frac{\phi(x)}{\phi(y_0)} .$$

It follows that

$$\phi(x) = \phi(y_0) (x , y_0) = (x , \overline{\phi(y_0)} \, y_0) .$$

Put $z = \overline{\phi(y_0)} \, y_0$. Then we have $\phi(x) = (x, z)$ for all $x \in H$.

To prove uniqueness, assume that we have two representations $\phi(x) = (x, v) = (x, w)$ for all $x \in H$. For $x = v - w$ we then have the computation

$$0 = (x , v) - (x , w) = (x , v - w) = (v - w , v - w) ,$$

which implies that $v - w = 0$, and hence $v = w$.

This completes the proof of Riesz' Representation Theorem. $\square$

3.6   Weak convergence

Definition 3.6.1. A sequence (x_n) in a Hilbert space H is said to be *weakly convergent* with *weak limit* x, if for all $y \in H$, the sequence of complex numbers

$((x_n, y))$ has the limit (x, y), in other words, if $(x_n, y) \to (x, y)$ for $n \to \infty$, for all $y \in H$.

If a sequence (x_n) in H converges weakly to $x \in H$, we write $x_n \rightharpoonup x$ for $n \to \infty$. In this context, ordinary convergence $x_n \to x$ for $n \to \infty$ is often called *strong* convergence.

Remark 3.6.2. Using Riesz' Representation Theorem, $x_n \rightharpoonup x$ for $n \to \infty$ if and only if $\phi(x_n) \to \phi(x)$ for $n \to \infty$ for every bounded linear functional ϕ on H. In this formulation, the notion of weak convergence can also be introduced in the more general setting of Banach spaces.

Lemma 3.6.3. *If $x_n \rightharpoonup x$ for $n \to \infty$, then the weak limit $x \in H$ is unique.*

Proof.   Suppose $x_n \rightharpoonup x$ and $x_n \rightharpoonup z$ for $n \to \infty$. Then $(x_n, y) \to (x, y)$ and $(x_n, y) \to (z, y)$ for $n \to \infty$, for all $y \in H$. Since limits are unique in $\mathbb{C}$, it follows that $(x, y) = (z, y)$ for all $y \in H$, and hence $x - z \in H^\perp = \{0\}$. $\square$

Example 3.6.4. Let (e_n) be an orthonormal sequence in the Hilbert space H. For $y \in H$, we have by Bessel's inequality, Theorem 3.4.5, that the series $\sum_{n=1}^{\infty} |(y, e_n)|^2$ is convergent and hence $(e_n, y) \to 0 = (0, y)$ for $n \to \infty$, for all $y \in H$. This shows that $e_n \rightharpoonup 0$ for $n \to \infty$.
On the other hand,

$$||e_n - e_m||^2 = ||e_n||^2 + ||e_m||^2 = 2 ,$$

for all $n \neq m$, so that (e_n) is not a Cauchy sequence, in particular, it is not strongly convergent. This justifies the name *weak convergence*.

Proposition 3.6.5. *Let (x_n) be a sequence in the Hilbert space H. If (x_n) converges strongly to x for $n \to \infty$, then (x_n) converges weakly to x for $n \to \infty$, in other words, if $x_n \to x$ for $n \to \infty$, then $x_n \rightharpoonup x$ for $n \to \infty$.*

Proof.   If $x_n \to x$ for $n \to \infty$, we have $||x_n - x|| \to 0$ for $n \to \infty$. From the Cauchy-Schwarz inequality follows

$$|(x_n, y) - (x, y)| = |(x_n - x, y)| \leq ||x_n - x|| \, ||y|| \quad \text{for all } y \in H .$$

This implies that $(x_n, y) \to (x, y)$ for $n \to \infty$, for all $y \in H$, proving that $x_n \rightharpoonup x$ for $n \to \infty$. $\square$

The difference between strong convergence and weak convergence is a genuine infinite dimensional phenomenon; if the Hilbert space is finite dimensional, weak convergence will imply strong convergence.

Theorem 3.6.6. *Let H be a finite dimensional Hilbert space. Then every weakly convergent sequence (x_n) in H is strongly convergent.*

Proof. Let $\{e_1, \ldots, e_k\}$ be an orthonormal basis for H. Then any vector $z \in H$ admits a unique decomposition

$$z = (z, e_1)e_1 + \cdots + (z, e_k)e_k .$$

Weak convergence of the sequence (x_n) to $x \in H$ implies $(x_n, e_i) \to (x, e_i)$ for $n \to \infty$, for all $i = 1, \ldots, k$. Hence

$$\|x - x_n\|^2 = \sum_{i=1}^{k} |(x, e_i) - (x_n, e_i)|^2 \to 0 \quad \text{for } n \to \infty .$$

This proves that $x_n \to x$ for $n \to \infty$ and hence completes the proof. $\square$

Theorem 3.6.7. *Every weakly convergent sequence in a Hilbert space H is bounded.*

Proof. Let (x_n) be a weakly convergent sequence in the Hilbert space H with weak limit $x \in H$. We have to prove that the sequence (x_n) is bounded, in other words, that the set of elements $\{x_n\}$ is a bounded subset of H.

We shall argue by contradiction. Hence we assume that the sequence (x_n) is unbounded. Clearly, the sequence $(x_n - x)$ is then also unbounded.

Let $K \in \mathbb{R}$ be a real number $K > 1$, which we shall specify later. We can then extract a subsequence (y_n) from the unbounded sequence $(x_n - x)$ such that $\|y_n\| \geq K^n$ for each $n \in \mathbb{N}$.

Now define the sequence (z_n) in H by a recursive procedure. For $n = 1$ put $z_1 = y_1$, and then for $n \geq 2$,

$$z_n = z_{n-1} + \begin{cases} K^{-n+1} \frac{(z_{n-1}, y_n)}{|(z_{n-1}, y_n)|} \frac{y_n}{\|y_n\|} & \text{if } (z_{n-1}, y_n) \neq 0 \\ K^{-n+1} \frac{y_n}{\|y_n\|} & \text{if } (z_{n-1}, y_n) = 0 . \end{cases}$$

Clearly, $\|z_n - z_{n-1}\| = K^{-n+1}$ for each $n \geq 2$. By the sum formula for a geometric series, it then follows that

$$\|z_m - z_n\| < \frac{K^{-n}}{1 - K^{-1}}$$

for all $m > n \geq 2$. This proves that (z_n) is a Cauchy sequence in the Hilbert space H. Consequently, (z_n) is convergent, say with limit $z \in H$.

First we consider the inner products (z_n, y_n) for $n \geq 2$.

If on the one hand $(z_{n-1}, y_n) \neq 0$, we have

$$
|(z_n, y_n)| = \left| \left(z_{n-1} + K^{-n+1} \frac{(z_{n-1}, y_n)}{|(z_{n-1}, y_n)|} \frac{y_n}{\|y_n\|}, y_n \right) \right|
$$

$$
= \left| (z_{n-1}, y_n) + (z_{n-1}, y_n) K^{-n+1} \frac{(y_n, y_n)}{|(z_{n-1}, y_n)| \|\|y_n\|} \right|
$$

$$
\geq K^{-n+1} \frac{|(z_{n-1}, y_n)| \|\|y_n\|^2}{|(z_{n-1}, y_n)| \|\|y_n\|} = \frac{\|y_n\|}{K^{n-1}} .
$$

If on the other hand $(z_{n-1}, y_n) = 0$, trivially

$$
|(z_n, y_n)| = \frac{\|y_n\|}{K^{n-1}} \quad \text{for each } n \geq 2 .
$$

Next we consider the inner products (z, y_n) for $n \geq 2$.

$$
|(z, y_n)| = |(z_n + z - z_n, y_n)| = |(z_n, y_n) + (z - z_n, y_n)|
$$

$$
\geq |(z_n, y_n)| - |(z - z_n, y_n)|
$$

$$
\geq |(z_n, y_n)| - \|z - z_n\| \|y_n\|
$$

$$
\geq \frac{\|y_n\|}{K^{n-1}} - \frac{K^{-n}}{1 - K^{-1}} \|y_n\|
$$

$$
= \frac{K(1 - K^{-1}) - 1}{(1 - K^{-1})} \frac{\|y_n\|}{K^n}
$$

$$
\geq K(1 - K^{-1}) - 1 = 1 \quad \text{if we put } K = 3.
$$

Now recall that (y_n) is a subsequence of $(x_n - x)$. Hence the above argument proves that there is a subsequence of $((z, x_n - x))$ that does not converge to 0, contradicting that the sequence (x_n) converges weakly to x.

This completes the proof of the theorem. □

Chapter 4

Operators on Hilbert Spaces

Applications of functional analysis often involve operations by continuous linear mappings on the elements in a normed vector space, in particular, a Hilbert space. The terminology used in this connection for such an operation is mostly that of a *bounded linear operator* rather than a *continuous linear mapping*.

For a normed vector space V, we denote by

$$B(V) \quad \text{the space of bounded linear operators} \quad T : V \to V \, ,$$

in other words, the space of continuous linear mappings $T : V \to V$.

4.1 The adjoint of a bounded linear operator

We open right away with the definition of the *adjoint* operator of a bounded linear operator on a Hilbert space and its operator norm properties.

Theorem 4.1.1. *Let H be a Hilbert space and $T \in B(H)$ a bounded linear operator on H. Then there is a unique bounded linear operator $T^* \in B(H)$ on H, called the* adjoint *operator of T, satisfying*

$$(Tx, y) = (x, T^*y) \quad for \ all \quad x, y \in H.$$

For the operator norms of T and T^*, it holds that

$$\|T^*\| = \|T\| \quad and \quad \|T^*T\| = \|T\|^2.$$

Furthermore, it holds that

$$(T^*)^* = T \quad and \quad (ST)^* = T^*S^*,$$

for all bounded linear operators $S, T \in B(H)$.

Proof. For every vector $y \in H$, we can define a bounded linear functional

$$\phi_y : H \to \mathbb{C} \quad \text{by} \quad \phi_y(x) = (Tx, y) \text{ for } x \in H .$$

The functional ϕ_y is indeed linear, since T is linear and the inner product is linear in the first argument. That ϕ_y is bounded follows by the computation, using Cauchy-Schwarz' inequality,

$$|\phi_y(x)| = |(Tx, y)| \leq \|Tx\| \|y\| \leq (\|T\| \|x\|) \|y\| = (\|T\| \|y\|) \|x\| .$$

By Riesz' Representation Theorem, there is a unique $z \in H$ such that

$$\phi_y(x) = (x, z) \quad \text{for all } x \in H .$$

To every $y \in H$, there is in other words a uniquely determined $z \in H$ for which

$$(Tx, y) = (x, z) \quad \text{for all } x \in H .$$

We can therefore define a map $T^* : H \to H$ by letting $T^*(y) = z$. The map T^* is characterized by the equation

$$(Tx, y) = (x, T^*y) \quad \text{for all } x, y \in H .$$

Assertion 4.1.2. T^* *is a linear operator.*

Proof. For arbitrary vectors $y_1, y_2 \in H$ and scalars $\alpha_1, \alpha_2 \in \mathbb{C}$, we have the following computation for every $x \in H$,

$$\begin{aligned}
(x, T^*(\alpha_1 y_1 + \alpha_2 y_2)) &= (Tx, \alpha_1 y_1 + \alpha_2 y_2) \\
&= \overline{\alpha_1}(Tx, y_1) + \overline{\alpha_2}(Tx, y_2) \\
&= \overline{\alpha_1}(x, T^*y_1) + \overline{\alpha_2}(x, T^*y_2) \\
&= (x, \alpha_1 T^*y_1 + \alpha_2 T^*y_2) .
\end{aligned}$$

Since this holds for all $x \in H$, we conclude that

$$T^*(\alpha_1 y_1 + \alpha_2 y_2) = \alpha_1 T^*y_1 + \alpha_2 T^*y_2 ,$$

completing the proof that T^* is linear. $\qquad\square$

Assertion 4.1.3. T^* *is bounded.*

Proof. For $y \in H$, we have the computation

$$\begin{aligned}
\|T^*y\|^2 &= (T^*y, T^*y) = (T(T^*y), y) \\
&\leq \|T(T^*y)\| \|y\| \leq \|T\| \|T^*y\| \|y\| .
\end{aligned}$$

From this follows that

$$||T^*y|| \le ||T||\,||y|| \quad \text{for all } y \in H .$$

[Trivial, if $||T^*y|| = 0$; by division, if $||T^*y|| \ne 0$.]

This proves that T^* is bounded, and furthermore, that $||T^*|| \le ||T||$. □

Assertion 4.1.4. $||T^*|| = ||T||$ *and* $||T^*T|| = ||T||^2$.

Proof. Interchanging the roles of T and T^* in the argument showing that $||T^*|| \le ||T||$, we get

$$||Tx||^2 = (Tx, Tx) = (x, T^*(Tx))$$
$$\le ||x||\,||T^*(Tx)|| \le ||x||\,||T^*||\,||Tx|| ,$$

from which follows,

$$||Tx|| \le ||T^*||\,||x|| \quad \text{for all } x \in H ,$$

proving that $||T|| \le ||T^*||$. Altogether, we conclude that $||T|| = ||T^*||$.

For $||x|| = 1$, we have the computation

$$||Tx||^2 = (Tx, Tx) = (x, T^*Tx) \le ||x||\,||T^*T||\,||x|| = ||T^*T|| ,$$

proving that $||T||^2 \le ||T^*T||$.

Since

$$||T^*T|| \le ||T^*||\,||T|| = ||T||\,||T|| = ||T||^2 ,$$

we get altogether $||T^*T|| = ||T||^2$, completing the proof of the assertion. □

Assertion 4.1.5. $(T^*)^* = T$ *and* $(ST)^* = T^*S^*$ *for any pair of bounded linear operators* $S, T \in B(H)$.

Proof. For all pairs of vectors $x, y \in H$, we have the computation

$$(Tx, y) = (x, T^*y) = \overline{(T^*y, x)} = \overline{(y, (T^*)^*x)} = ((T^*)^*x, y) ,$$

from which follows that $Tx = (T^*)^*x$ for all $x \in H$, proving that $(T^*)^* = T$.

The computation

$$(x, (ST)^*y) = (STx, y) = (Tx, S^*y) = (x, T^*S^*y) ,$$

valid for all $x, y \in H$, proves that

$$(ST)^*y = T^*S^*y \quad \text{for all } y \in H ,$$

or, in other words, that $(ST)^* = T^*S^*$. □

This completes the proof of Theorem 4.1.1.   □

Using Theorem 3.5.9 we can prove the following theorem.

Theorem 4.1.6. *Every bounded linear operator $T \in B(H)$ on a Hilbert space H induces the orthogonal splitting*

$$H = \ker(T^*) \oplus \overline{T(H)} \; ,$$

where $\ker(T^*)$ denotes the kernel of T^* and $\overline{T(H)}$ is the closure of the image $T(H)$ for T.

Proof. By the Projection Theorem,

$$H = \overline{T(H)}^{\perp} \oplus \overline{T(H)} \; ,$$

valid for every closed linear subspace in H, in particular $\overline{T(H)}$.

Assertion 4.1.7. $\overline{T(H)}^{\perp} = T(H)^{\perp} = \ker(T^*)$.

Proof. Let $y \in T(H)^{\perp}$. Then

$$(x, T^*y) = (Tx, y) = 0 \quad \text{for all } x \in H \; ,$$

proving that $T^*y = 0$, i.e. $y \in \ker(T^*)$.

On the other hand, let $y \in \ker(T^*)$. Then

$$(Tx, y) = (x, T^*y) = (x, 0) = 0 \quad \text{for all } x \in H \; ,$$

proving that $y \in T(H)^{\perp}$.

Altogether, $T(H)^{\perp} = \ker(T^*)$. It is easy to prove that $\overline{T(H)}^{\perp} = T(H)^{\perp}$.

This completes the proof of the assertion and hence of the theorem.   □

Example 4.1.8. A bounded linear operator $T : \mathbb{C}^k \to \mathbb{C}^k$ is represented by a $k \times k$-matrix $[T_{ij}]$ with respect to the canonical basis $\{e_1, \ldots, e_k\}$ in $\mathbb{C}^k$. In terms of the inner product in $\mathbb{C}^k$, the matrix elements are determined by $T_{ij} = (Te_j, e_i)$.

Let $[T^*_{ij}]$ be the matrix representing the adjoint operator $T^* : \mathbb{C}^k \to \mathbb{C}^k$. Then

$$T^*_{ij} = (T^*e_j, e_i) = \overline{(e_i, T^*e_j)} = \overline{(Te_i, e_j)} = \overline{T_{ji}} \; .$$

Conclusion: *The matrix representation for the adjoint linear operator T^* is the conjugate transpose of the matrix representation for T.*

In the case of real vector spaces, the adjoint operator $T^* : \mathbb{R}^k \to \mathbb{R}^k$ for a bounded linear operator $T : \mathbb{R}^k \to \mathbb{R}^k$ is represented by the transposed matrix to the matrix representing T.

Definition 4.1.9. Let H be a Hilbert space. A bounded linear operator T on H is called *self-adjoint* if $T = T^*$.

Lemma 4.1.10. *Let T be a bounded, self-adjoint linear operator on a Hilbert space H. Then (Tx, x) is a real number for all $x \in H$.*

Proof. For every $x \in H$, we have the computation

$$(Tx, x) = (x, T^*x) = (x, Tx) = \overline{(Tx, x)} \ ,$$

proving that (Tx, x) is a real number. □

For self-adjoint operators, there is an alternative method to determine the operator norm.

Theorem 4.1.11. *Let T be a bounded, self-adjoint linear operator on a Hilbert space H. Then*

$$\|T\| = \sup\{\,|(Tx, x)|\,\big|\, x \in H, \|x\| = 1\} \ .$$

Proof. For each vector $x \in H$ with $\|x\| = 1$, we get by the Cauchy-Schwarz inequality,

$$|(Tx, x)| \leq \|Tx\|\|x\| \leq \|T\|\|x\|^2 = \|T\| \ .$$

This proves that the number

$$M_T = \sup\{\,|(Tx, x)|\,\big|\, x \in H, \|x\| = 1\}$$

exists, and that $M_T \leq \|T\|$.

A simple calculation shows that

$$(T(x + y), x + y) - (T(x - y), x - y) = 2(Tx, y) + 2(Ty, x) \ ,$$

where each term on the left-hand side is a real number since T is self-adjoint.

From the definition of M_T follows that

$$(T(x + y), x + y) \leq M_T \|x + y\|^2 \ ,$$
$$-(T(x - y), x - y) \leq M_T \|x - y\|^2 \ .$$

To prove these inequalities, write $x + y = \|x + y\|u$ and $x - y = \|x - y\|v$ for unit vectors $u, v \in H$.

Using the parallelogram law, Proposition 3.5.1, for the second inequality
we get,

$$(Tx, y) + (Ty, x) = \frac{1}{2}\left((T(x+y), x+y) - (T(x-y), x-y)\right)$$

$$\leq \frac{1}{2} M_T\left(||x+y||^2 + ||x-y||^2\right)$$

$$\leq M_T\left(||x||^2 + ||y||^2\right).$$

Assertion 4.1.12. $||Tx|| \leq M_T ||x||$ *for all* $x \in H$.

Proof.   The assertion is trivial for $Tx = 0$. So assume that $Tx \neq 0$. Put

$$y = \frac{||x||}{||Tx||} Tx.$$

Then we have the computations,

$$(Tx, y) + (Ty, x) = \frac{||x||}{||Tx||}(Tx, Tx) + \frac{||x||}{||Tx||}(T(Tx), x)$$

$$= ||x|| ||Tx|| + \frac{||x||}{||Tx||}(Tx, T^*x)$$

$$= ||x|| ||Tx|| + \frac{||x||}{||Tx||}(Tx, Tx) \quad (T^* = T)$$

$$= 2 ||x|| ||Tx||.$$

$$M_T\left(||x||^2 + ||y||^2\right) = M_T\left(||x||^2 + \frac{||x||^2}{||Tx||^2}||Tx||^2\right) = 2 M_T ||x||^2.$$

By the inequality preceding the assertion we conclude that

$$||Tx|| \leq M_T ||x|| \quad \text{for all } x \in H,$$

as asserted. $\square$

From the assertion we infer that $||T|| \leq M_T$. Together with the already
established inequality $M_T \leq ||T||$ this proves that $||T|| = M_T$, and the proof
of the theorem is complete. $\square$

In finite dimensions we know that a linear operator $T : \mathbb{C}^k \to \mathbb{C}^k$ is injective if and only if it is surjective. For self-adjoint linear operators we have a
corresponding result in the Hilbert space case, valid also in infinite dimensions.

Theorem 4.1.13. *Let T be a bounded, self-adjoint linear operator on a Hilbert space H. If $T(H)$ is dense in H, then T has an inverse operator defined on the image $T(H)$ of T.*

Proof. By Theorem 4.1.6 we get the splitting $H = \ker(T^*) \oplus \overline{T(H)}$ from which follows that $\ker(T^*) = \{0\}$, since $T(H)$ is dense in H. But then $\ker(T) = \{0\}$, since $T^* = T$, proving that T is injective. Altogether, we have a well-defined inverse linear operator $T^{-1} : T(H) \to H$ as stated in the theorem. $\square$

Remark 4.1.14. The inverse linear operator T^{-1} in Theorem 4.1.13 is not necessarily bounded; in fact, most often in applications it is not! If T^{-1} is unbounded, its domain of definition cannot be extended to all of H, since it is equivalent for T^{-1} to be bounded and to admit an extension from the dense subspace $T(H)$ to all of H. That these statements are equivalent follows since the inverse mapping for a bijective, bounded linear mapping between Banach spaces is always bounded. The latter fact is a corollary to the famous Open Mapping Theorem known as Banach's Theorem 1.8.3.

Theorem 4.1.15. *Let P be a bounded, self-adjoint linear operator on the Hilbert space H, and assume that P is an* idempotent *operator, i.e. $P^2 = P$.*
 Then P is the orthogonal projection onto the closed linear subspace

$$M = \{z \in H \mid Pz = z\} \quad (fixed \ space \ for \ P).$$

Proof. Let $I : H \to H$ denote the identity operator on H, i.e. $I(z) = z$ for all $z \in H$. Then

$$M = \{z \in H \mid Pz = z\} = \{z \in H \mid (P - I)z = 0\} = \ker(P - I) \ .$$

Since $P - I$ is a continuous linear operator on H, it follows from the description of M as the kernel of $P - I$ that M is a closed linear subspace in H.

By Theorem 3.5.9, we then have the orthogonal splitting

$$H = M \oplus M^\perp \ .$$

Hence every vector $z \in H$ admits a unique decomposition,

$$z = x + y, \ x \in M, y \in M^\perp \ .$$

From the decomposition $z = x + y$ with $x \in M$ and $y \in M^\perp$, we get $Pz = Px + Py = x + Py$. Since $P(Py) = P^2 y = Py$, we have $Py \in M$. But Py also belongs to $M^\perp$. This follows from the computation $(x, Py) = (Px, y) = (x, y) = 0$, valid for all $x \in M$. Since $M \cap M^\perp = \{0\}$, we conclude that $Py = 0$.

For $z = x + y$, $x \in M, y \in M^{\perp}$, it follows altogether that $Pz = x$. In other words, P is exactly the orthogonal projection onto the fixed space M for P, and the proof is complete. $\square$

For obvious reasons, a bounded, self-adjoint, idempotent linear operator P on a Hilbert space H is called a *projection operator*. As we have seen in Theorem 4.1.15, a projection operator is exactly the orthogonal projection onto the corresponding fixed space.

Proposition 4.1.16. *For a projection operator P on a Hilbert space H, the operator norm satisfies $\|P\| \leq 1$. If $P \neq 0$, the operator norm $\|P\| = 1$.*

Proof. For all $x \in H$ we have the computation

$$\|Px\|^2 = (Px, Px) = (P^2 x, x) = (Px, x) \leq \|Px\| \|x\|,$$

proving that $\|Px\| \leq 1$ for all $x \in H$ with $\|x\| = 1$. It follows that $\|P\| \leq 1$.

If $P \neq 0$, the fixed space $M = \{z \in H \mid Pz = z\}$ for P contains nonzero vectors [with $z = Pu$, we have $Pz = P(Pu) = P^2 u = Pu = z$]. For a unit vector $x \in M$, we have $\|Px\| = \|x\| = 1$, and consequently, the operator norm must be $\|P\| = 1$. $\square$

4.2 Compact operators

In a metric space, in particular in a normed vector space, the notions of compactness (defined by the covering property) and sequential compactness (every sequence contains a convergent subsequence) are equivalent notions. When the notion of compactness is used in normed vector spaces, we are therefore free to use the definition most suitable for our purposes. In functional analysis it is often convenient to use the definition: A subset K in a normed vector space V is *compact*, if every sequence (x_n) of elements in K contains a convergent subsequence (x_{n_k}) with limit point $x \in K$.

In a finite dimensional normed vector space V we know by the Heine-Borel Theorem that a subset $K \subseteq V$ is compact if and only if it is a closed and bounded subset.

It is always true that a compact subset $K \subseteq V$ of a normed vector space V is closed and bounded. In general, the converse is only true in finite dimensions.

Example 4.2.1. Let H be a separable Hilbert space. Consider the unit ball

$$C = \{x \in H \mid \|x\| \leq 1\}.$$

Clearly, C is a closed and bounded subset of H. But C is not compact if H is infinite dimensional. This follows since an arbitrary orthonormal sequence (x_n) in H is contained in C and has no convergent subsequence (x_{n_k}); a subsequence is not even a Cauchy sequence, since $||x_{n_i} - x_{n_j}||^2 = 2$ for all $i \neq j$.

The result in Example 4.2.1 holds in complete generality in a normed vector space. To prove this we need a famous result known as Riesz' Lemma.

Lemma 4.2.2 (Riesz' Lemma). *Let V be a normed vector space and let U be a closed subspace of V with $U \neq V$. Let α be a real number in the interval $0 < \alpha < 1$. Then there exists a vector $v \in V$ such that*

$$||v|| = 1 \quad and \quad ||v - u|| \geq \alpha \quad for\ all\ u \in U .$$

Proof. Choose an arbitrary vector $v_0 \in V \setminus U$. Since U is a closed subspace of V, the distance of v_0 to U is strictly positive, i.e.

$$\inf\{||v_0 - u|| \,|\, u \in U\} = r_0 > 0 .$$

Since $r_0 < r_0/\alpha$, when $0 < \alpha < 1$, we can choose $u_0 \in U$ such that

$$||v_0 - u_0|| \leq \frac{r_0}{\alpha} .$$

Put $v = (v_0 - u_0)/||v_0 - u_0||$. Then $||v|| = 1$ and

$$||v - u|| = \frac{||v_0 - u_0 - ||v_0 - u_0||\, u||}{||v_0 - u_0||} \geq \frac{r_0}{||v_0 - u_0||} \geq \alpha ,$$

for all $u \in U$. This completes the proof. □

Theorem 4.2.3. *Let V be an arbitrary normed vector space, and let*

$$C = \{\, x \in V \,|\, ||x|| \leq 1 \}$$

he the closed unit ball in V.

Then C is compact if and only if V has finite dimension.

Proof. First assume that C is compact and that V is not finite dimensional. Choose an arbitrary element $x_1 \in V$ with $||x_1|| = 1$. Let U_1 be the 1-dimensional, and hence closed, subspace of V spanned by x_1. By Riesz' Lemma 4.2.2, there exists $x_2 \in V \setminus U_1$ with $||x_2|| = 1$ and $||x_2 - x_1|| \geq 1/2$. Let U_2 be the 2-dimensional, and hence closed, subspace of V spanned by $\{x_1, x_2\}$. Again by Riesz' Lemma, there exists $x_3 \in V \setminus U_2$ with $||x_3|| = 1$ and $||x_3 - x_1|| \geq 1/2$, $||x_3 - x_2|| \geq 1/2$. Continuing by induction, we find a sequence (x_n) of linearly independent vectors in V with the properties $||x_n|| = 1$ and $||x_n - x_m|| \geq 1/2$, for $n \neq m$. By construction, the sequence (x_n) in C does not contain a convergent subsequence. This contradicts that C is compact. Therefore necessarily V has finite dimension if C is compact.

Conversely, if V is finite dimensional, then C is compact by the Heine-Borel Theorem 1.3.12, since it is a closed and bounded subset in V. □

A linear operator between normed vector spaces is by definition a *bounded*, or, *continuous*, operator if it maps bounded subsets in the domain vector space into bounded subsets in the image vector space. A *compact*, or, *completely continuous*, operator has a stronger property.

Definition 4.2.4. Let V and W be normed vector spaces. A bounded linear operator $T : V \to W$ is said to be a *compact*, or *completely continuous*, operator if it maps every bounded subset A in V into a subset $T(A)$ in W with compact closure $\overline{T(A)}$ in W.

By Example 4.2.1, the unit ball in a Hilbert space H is not compact in infinite dimension. Hence the identity operator I on H is not compact. Somehow, compactness of an operator is related to finite dimensionality of its image.

Definition 4.2.5. A bounded linear operator $T : V \to W$ between normed vector spaces V and W is said to have *finite rank* if $\dim T(V)$ is finite, where $\dim T(V)$ is the dimension of the linear subspace in W formed by the image for T in W.

Proposition 4.2.6. *A bounded linear operator of finite rank is a compact operator.*

Proof. Let $T : V \to W$ be a bounded linear operator between normed vector spaces V and W for which $\dim T(V)$ is finite. Since T is bounded, it maps a bounded subset $A \subseteq V$ into a bounded subset $T(A) \subseteq T(V) \subseteq W$. Also the closure $\overline{T(A)}$ is contained in $T(V)$, since $T(V)$ is finite dimensional and hence a closed linear subspace in W. Consequently, $\overline{T(A)}$ is bounded and closed in a finite dimensional vector space and therefore compact. □

In finite dimensions, weak convergence coincides with convergence in norm and hence it is to be expected that compact linear operators behave well with respect to weak convergence. This is indeed true, and the following result motivates why compact linear operators are also called completely continuous operators.

Theorem 4.2.7. *Let H be a Hilbert space and let (x_n) be a weakly convergent sequence with weak limit x. Then for every compact linear operator T on H,*

*the image sequence (Tx_n) converges strongly (in norm) to Tx. In other words:
For a compact linear operator $T \in B(H)$ it holds that*

$$x_n \rightharpoonup x \ for \ n \to \infty \ \Rightarrow \ Tx_n \to Tx \ for \ n \to \infty \ .$$

Proof. If $x_n \rightharpoonup x$ for $n \to \infty$ we have for all $y \in H$,

$$(Tx_n, y) = (x_n, T^*y) \to (x, T^*y) = (Tx, y) \quad for \quad n \to \infty \ .$$

This proves that $Tx_n \rightharpoonup Tx$ for $n \to \infty$. Since weak limits are unique and
strong convergence implies weak convergence, the only possible strong limit
for (Tx_n) is Tx.

Assertion 4.2.8. $Tx_n \to Tx$ *for $n \to \infty$.*

Proof. The proof is indirect. Suppose (Tx_n) does not converge in norm to
Tx. Then we can extract a subsequence (Tx_{n_k}) of (Tx_n) such that

$$\|Tx_{n_k} - Tx\| \geq \delta \quad for \ all \ k \in \mathbb{N} \quad and \ some \ \delta > 0 \ .$$

Since (x_n) is bounded by Theorem 3.6.7 and T is a compact linear operator,
we can extract a subsequence $(Tx_{n_{k_l}})$ of (Tx_{n_k}) for which $Tx_{n_{k_l}} \to y \in H$ for
$l \to \infty$. Since $Tx_{n_{k_l}} \rightharpoonup Tx$ for $l \to \infty$, we must have $y = Tx$, which is not
possible by the construction of Tx_{n_k}. Thereby we have obtained a contradiction
and hence $Tx_n \to Tx$ for $n \to \infty$. □

This completes the proof of the theorem. □

Example 4.2.9. Let (e_n) be an orthonormal sequence in the Hilbert space
H. From Example 3.6.4 we know that $e_n \rightharpoonup 0$ for $n \to \infty$. Hence by Theorem
4.2.7 we conclude that $Te_n \to 0$ for $n \to \infty$ for every compact linear operator
$T \in B(H)$.

From the example we infer that the inverse operator (if it exists) of a
compact linear operator on a separable Hilbert space is always unbounded.

Proposition 4.2.10. *Let H be an infinite dimensional, separable Hilbert
space. Suppose $T \in B(H)$ is a compact linear operator on H, which has an
inverse T^{-1} on $T(H)$. Then T^{-1} is always an unbounded linear operator.*

Proof. Let (e_n) be an orthonormal sequence in H; for example an orthonormal basis in H. Then $Te_n \to 0$ for $n \to \infty$, but on the other hand,
$\|T^{-1}(Te_n)\| = \|e_n\| = 1$ for all $n \in \mathbb{N}$, and hence T^{-1} cannot be continuous.
In other words, the linear operator T^{-1} is unbounded. □

The set of compact linear operators on a Hilbert space H is a closed subset in the Banach space (operator norm) of bounded linear operators $B(H)$ on H. This is a consequence of the following theorem.

Theorem 4.2.11. *Let H be a Hilbert space and suppose that (T_n) is a sequence of compact linear operators in $B(H)$ converging (in operator norm) to the bounded linear operator $T \in B(H)$. Then T is a compact linear operator.*

Proof.   It is sufficient to prove that if (x_n) is a bounded sequence in H, then it is possible to extract a subsequence (x_{n_k}) such that (Tx_{n_k}) is convergent.

Let (x_n) be a bounded sequence in H. Since T_1 is compact, (x_n) has a subsequence (x_n^1) such that $(T_1x_n^1)$ is convergent. Since T_2 is compact, (x_n^1) has a subsequence (x_n^2) such that $(T_2x_n^2)$ is convergent. Continuing this way we get for every $k \in \mathbb{N}$, a subsequence (x_n^k) of (x_n) such that $(T_kx_n^k)$ is convergent. Now consider the *diagonal sequence* (x_n^n). It is obvious that $(T_kx_n^n)$ is convergent for every $k \in \mathbb{N}$.

Assertion 4.2.12. *(Tx_n^n) is a Cauchy sequence in H.*

Proof.   Given $\varepsilon > 0$.

Since the original sequence (x_n) is bounded, there exists a positive constant C such that $\|x_n^n\| \leq C$ for all $n \in \mathbb{N}$.

Since $T_n \to T$ for $n \to \infty$ (in operator norm), we can choose a number $k \in \mathbb{N}$, which we then keep fixed, such that

$$\|T - T_k\| \leq \frac{\varepsilon}{3C} .$$

Since $(T_kx_n^n)$ is convergent, there exists a number $n_0 \in \mathbb{N}$, such that

$$\|T_kx_n^n - T_kx_m^m\| \leq \frac{\varepsilon}{3} \quad \text{for all} \quad n, m \geq n_0 .$$

For $n, m \geq n_0$ we now have the computation

$$\|Tx_n^n - Tx_m^m\| \leq \|Tx_n^n - T_kx_n^n\| + \|T_kx_n^n - T_kx_m^m\| + \|T_kx_m^m - Tx_m^m\|$$
$$\leq \|T - T_k\|\|x_n^n\| + \frac{\varepsilon}{3} + \|T_k - T\|\|x_m^m\|$$
$$\leq \frac{\varepsilon}{3C}C + \frac{\varepsilon}{3} + \frac{\varepsilon}{3C}C = \varepsilon .$$

This proves that (Tx_n^n) is indeed a Cauchy sequence in H.   □

By completeness of H, the Cauchy sequence (Tx_n^n) is convergent in H. Since (x_n^n) is a subsequence of (x_n), this proves that T is a compact linear operator.   □

The following theorem provides among others a method for constructing compact linear operators.

Theorem 4.2.13. *Let (e_n) be an orthonormal basis for the infinite dimensional, separable Hilbert space H and let (λ_n) be an arbitrary sequence of complex numbers. For every $x \in H$, consider the infinite series*

$$Tx = \sum_{n=1}^{\infty} \lambda_n (x, e_n) e_n .$$

(1) *The infinite series is convergent and the sum defines a linear operator T on H if the sequence (λ_n) is bounded.*

(2) *T exists and is bounded if and only if the sequence (λ_n) is bounded.*

(3) *T exists and is a compact linear operator if and only if $\lambda_n \to 0$ for $n \to \infty$.*

Proof.   The series $\sum_{n=1}^{\infty} (x, e_n) e_n$ is convergent for all $x \in H$ by Theorem 3.4.5 (Bessel's inequality). Then by the comparison test, the infinite series $Tx = \sum_{n=1}^{\infty} \lambda_n (x, e_n) e_n$ is also convergent for all $x \in H$ if the sequence (λ_n) is bounded, in particular, if $\lambda_n \to 0$ for $n \to \infty$. In case of existence, clearly T is a linear operator. This proves (1).

Since

$$\|Tx\|^2 = \sum_{n=1}^{\infty} |\lambda_n|^2 |(x, e_n)|^2 \quad \text{and} \quad \|x\|^2 = \sum_{n=1}^{\infty} |(x, e_n)|^2 ,$$

and since $\|Te_n\| = |\lambda_n|$ for all $n \in \mathbb{N}$, it follows immediately that T exists and is bounded if and only if the sequence (λ_n) is bounded. This proves (2).

In order to prove (3), we define the sequence of operators (T_k) by

$$T_k x = \sum_{n=1}^{k} \lambda_n (x, e_n) e_n \quad \text{for all } x \in H .$$

Since T_k has finite rank, all the operators T_k are compact linear operators by Proposition 4.2.6.

When the sequence (λ_n) is bounded, we can define the bounds

$$K_k = \sup_{n>k} \{|\lambda_n|^2\} \quad \text{for all } k \in \mathbb{N} .$$

Then for all $k \in \mathbb{N}$, we get the estimate

$$\|Tx - T_k x\|^2 = \sum_{n=k+1}^{\infty} |\lambda_n|^2 |(x, e_n)|^2 \le K_k \|x\|^2 .$$

From this estimate follows that $||T - T_k|| \leq \sqrt{K_k}$ and hence that

$$||T - T_k|| \to 0 \quad \text{for } k \to \infty, \quad \text{if} \quad \lambda_n \to 0 \quad \text{for } n \to \infty .$$

This proves that $T_k \to T$ for $k \to \infty$, and hence by Theorem 4.2.11 that T is a compact linear operator, if $\lambda_n \to 0$ for $n \to \infty$.

Now assume that the sequence (λ_n) does not converge to 0 for $n \to \infty$. Then there exist an $\varepsilon > 0$ and a subsequence (λ_{n_k}) of (λ_n) such that $|\lambda_{n_k}| \geq \varepsilon$ for all $k \in \mathbb{N}$. Consider the corresponding subsequence (e_{n_k}) of (e_n). Being an orthonormal sequence, (e_{n_k}) is weakly convergent to 0. Since

$$||Te_{n_i} - Te_{n_j}||^2 = ||\lambda_{n_i} e_{n_i} - \lambda_{n_j} e_{n_j}||^2 = |\lambda_{n_i}|^2 + |\lambda_{n_j}|^2 \geq 2\varepsilon^2 ,$$

for all $i \neq j$, the sequence (Te_{n_k}) can have no subsequences that are Cauchy sequences, and hence in particular, (Te_{n_k}) does not converge to 0. From this we conclude by Theorem 4.2.7 that T is not a compact linear operator.

This completes the proof of the theorem. $\qquad \square$

Example 4.2.14. Let (e_n) be an orthonormal basis in the infinite dimensional, separable Hilbert space H. Define

$$Tx = \sum_{n=1}^{\infty} \frac{1}{n} (x, e_n) e_n \quad \text{for all } x \in H .$$

Then T is a compact linear operator on H by Theorem 4.2.13. The following computation shows that the operator T is self-adjoint,

$$(Tx, y) = \sum_{n=1}^{\infty} \frac{1}{n} (x, e_n) \overline{(y, e_n)} = (x, Ty) \quad \text{for all } x, y \in H .$$

Since $T(ke_k) = e_k$ for all $k \in \mathbb{N}$, it follows that $T(H)$ is dense in H. Hence T has an inverse operator $T^{-1} : T(H) \to H$ by Theorem 4.1.13. Since $T^{-1}(e_k) = ke_k$ for all $k \in \mathbb{N}$, clearly T^{-1} is unbounded.

Chapter 5

Spectral Theory

A linear map $T : \mathbb{C}^k \to \mathbb{C}^k$ is represented by a complex $k \times k$-matrix with respect to the canonical basis in $\mathbb{C}^k$. If the linear map T is self-adjoint, we know from Example 4.1.8 that it is represented by a conjugate symmetric matrix. From basic linear algebra, it is then well known that there exists an orthonormal basis $\{e_j\}_{j=1}^k$ for $\mathbb{C}^k$ in which the matrix for T can be diagonalized, or, which amounts to the same thing, that the linear map itself is given by

$$ Tx = \sum_{j=1}^{k} \lambda_j (x, e_j) e_j, \quad \text{for all } x \in \mathbb{C}^k . $$

Here (x, e_j) is the jth coordinate of x in an orthonormal basis $\{e_j\}_{j=1}^k$ of eigenvectors for T corresponding to the eigenvalues $\{\lambda_j\}_{j=1}^k$ for T.

The eigenvalues for T are the complex numbers λ, for which the equation $Tx = \lambda x$ has nontrivial solutions $x \in \mathbb{C}^k$, or, by an equivalent formulation, for which the linear operator $T - \lambda I$ is not injective, where I denotes the identity operator on $\mathbb{C}^k$.

In infinite dimensions, the situation is much more complicated, but for self-adjoint, compact linear operators a corresponding theory can be developed. It culminates in the famous *spectral theorem* - the main goal of this chapter.

5.1 The spectrum and the resolvent

Let H be a Hilbert space and let $T : H \to H$ be a linear operator. Often in the following, a linear operator T may not be defined on all of H but only on a linear subspace of H called the *domain of definition* and denoted by $D(T)$.

Let $T : H \to H$ be a linear operator with domain of definition $D(T) \subseteq H$.

For $\lambda \in \mathbb{C}$, we define the linear operator

$$T_\lambda = T - \lambda I \quad \text{with } D(T_\lambda) = D(T) ,$$

where I denotes the identity operator on H.

Remark 5.1.1. Integral equations of the form

$$(T - \lambda I)u = f \quad \text{for} \quad u, f \in L^2([a,b]), \ \lambda \in \mathbb{C} ,$$

defined in the Hilbert space $H = L^2([a,b])$, are known under the name of *Fredholm integral equations* of the second kind. They occur frequently in applications in the physical sciences.

In the general setting presented above, we consider the equation

$$T_\lambda x = y \quad \text{for each } y \in H .$$

If T_λ is injective, we can solve the equation to get

$$x = T_\lambda^{-1} y \quad \text{for all } y \in T_\lambda(H) = D(T_\lambda^{-1}) .$$

If T_λ^{-1} is also bounded, and $T_\lambda(H) = D(T_\lambda^{-1})$ is dense in H, then T_λ^{-1} admits a unique extension to all of H, such that the equation $T_\lambda x = y$ can be solved for all $y \in H$. This is the ideal situation from the point of view of the equation. However, it will not always be the case.

The operator

$$R_\lambda(T) = T_\lambda^{-1} = (T - \lambda I)^{-1} ,$$

defined for those $\lambda \in \mathbb{C}$ where T_λ is injective, is called the *resolvent* of T. We are particularly interested in the situation where $D(R_\lambda(T)) = T_\lambda(H)$ is dense in H and $R_\lambda(T)$ is a bounded linear operator.

Definition 5.1.2. The *resolvent set* for T, denoted by $\rho(T)$, is the set of $\lambda \in \mathbb{C}$ for which $R_\lambda(T)$ exists as a densely defined and bounded linear operator on H. The complement $\sigma(T) = \mathbb{C} \setminus \rho(T)$ is called the *spectrum* for T.

For $\lambda \in \sigma(T)$ various things can happen. First of all, maybe the linear map T_λ is not injective so that the resolvent $R_\lambda(T)$ is not defined. This is the case if and only if the equation

$$(T - \lambda I)x = 0 \quad \text{has a nontrivial solution} \quad x \in H ,$$

or equivalently,

$$Tx = \lambda x \quad \text{for a nontrivial vector} \quad x \in H .$$

If this is the case, we say that λ is an *eigenvalue* of T, with corresponding *eigenvector* $x \in H$. The equation $Tx = \lambda x$ is always satisfied for $x = 0$ and hence we emphasize that in order for λ to be an eigenvalue for T, the equation shall be satisfied for an eigenvector $x \neq 0$. Furthermore, we notice that the collection of all eigenvectors, belonging to an eigenvalue λ, form a linear subspace H_λ of H, called the *eigenspace* corresponding to λ. The linear subspace H_λ can be of finite or infinite dimension depending on the case in question.

Definition 5.1.3. The subset $\sigma_P(T)$ of the spectrum $\sigma(T)$ for T consisting of the eigenvalues for T is called the *point spectrum* for T.

The subset $\sigma_C(T)$ of the spectrum $\sigma(T)$ for T consisting of the complex numbers $\lambda \in \mathbb{C}$ for which $R_\lambda(T)$ exists as a densely defined but unbounded operator on H, is called the *continuous spectrum* for T.

Finally, the subset $\sigma_r(T)$ of the spectrum $\sigma(T)$ for T, where $R_\lambda(T)$ exists but is not densely defined, is called the *residual spectrum* for T.

All possibilities for $\lambda \in \sigma(T)$ are covered by Definition 5.1.3, and hence the spectrum $\sigma(T)$ for T is the disjoint union of the point spectrum, the continuous spectrum and the residual spectrum for T. In concrete cases, the resolvent set $\rho(T)$ and/or parts of the spectrum for T may be empty.

Example 5.1.4. Let $T : \mathbb{C}^k \to \mathbb{C}^k$ be a linear map with corresponding matrix $\mathbf{T} = [T_{ij}]$. Then it is well known that T has k eigenvalues $\lambda_1, \ldots, \lambda_k$ (counted with multiplicity) and that the eigenvalues are the roots of the characteristic polynomial $P(\lambda) = \det(\mathbf{T} - \lambda \mathbf{I})$ for T. Since $R_\lambda(T) \in B(\mathbb{C}^k)$, when λ is not an eigenvalue, it follows that the spectrum $\sigma(T)$ is a pure point spectrum, namely $\sigma(T) = \{\lambda_j\}_{j=1}^k$, and that the resolvent set is the complex plane except finitely many points, namely $\rho(T) = \mathbb{C} \setminus \{\lambda_j\}_{j=1}^k$.

Example 5.1.5. Let $\partial = d/dx$ be the differentiation operator with domain of definition $D(\partial) = C^1([0,1])$ in the Hilbert space $H = L^2([0,1])$. Then

$$(\partial - \lambda I)e^{\lambda x} = 0 \quad \text{for all } \lambda \in \mathbb{C} .$$

It follows that the spectrum $\sigma(\partial)$ is a pure point spectrum, namely $\sigma(\partial) = \mathbb{C}$, and consequently, that the resolvent set $\rho(\partial) = \mathbb{C} \setminus \sigma(\partial) = \emptyset$.

The unbounded linear operator ∂ in Example 5.1.5 has an empty resolvent set. On the other hand, this is never the case for a bounded linear operator. This will be an immediate consequence of interesting topological properties of the spectrum for a bounded linear operator. For the proof of these properties, we need a general result on existence of bounded inverse operators for small

perturbations of an identity operator proved by the German mathematician
Carl Gottfried Neumann (1832–1925).

Lemma 5.1.6 (Neumann's Lemma). *Let V be an arbitrary Banach space, and
let $T \in B(V)$ be a bounded linear operator with operator norm $||T|| < 1$. Then
the operator $I - T$ has a bounded inverse operator $(I - T)^{-1}$ on V.*

Proof.   Consider the infinite series,

$$S = I + \sum_{n=1}^{\infty} T^n = I + T + T^2 + \cdots + T^n + \dots \ ,$$

in the Banach space $B(V)$. The series is convergent in $B(V)$ by the comparison
test, since

$$||T^n|| \leq ||T||^n \quad \text{for all } n \in \mathbb{N} \ ,$$

and the geometric series $\sum_{n=0}^{\infty} ||T||^n$ is convergent, since $||T|| < 1$.
    Elementary computations show that

$$S(I - T) = I = (I - T)S \ .$$

This proves that $I - T$ is invertible with inverse operator $(I - T)^{-1} = S$.   □

    Then we are ready to study the topological properties of the resolvent set
$\rho(T)$ and the spectrum $\sigma(T)$ for a bounded linear operator T on a Hilbert
space H.

Theorem 5.1.7. *Let H be a Hilbert space and let $T \in B(H)$ be a bounded
linear operator on H. Then the resolvent set $\rho(T)$ for T is an open set in $\mathbb{C}$
and the spectrum $\sigma(T)$ for T is a closed set in $\mathbb{C}$.*

Proof.   If $\rho(T) = \emptyset$, the theorem is trivially true.
    Assume therefore that $\rho(T) \neq \emptyset$. Let $\mu \in \rho(T)$ be an arbitrary point in
$\rho(T)$. Then $R_\mu(T) = (T - \mu I)^{-1}$ is a densely defined, bounded linear operator.
For $\lambda \in \mathbb{C}$, we have

$$T - \lambda I = (T - \mu I) - (\lambda - \mu)I = (T - \mu I)\big(I - (\lambda - \mu)R_\mu(T)\big) \ .$$

By Neumann's Lemma, this rewriting shows that $T - \lambda I$ is an invertible, densely
defined (namely on $T_\mu(H)$) operator with bounded inverse, when

$$||(\lambda - \mu)R_\mu(T)|| = |\lambda - \mu|\,||R_\mu(T)|| < 1 \ ,$$

in other words, for $|\lambda - \mu| < ||R_\mu(T)||^{-1}$. This describes a circle in $\mathbb{C}$ with
center μ, thereby proving that $\rho(T)$ is an open set in $\mathbb{C}$.

Since $\sigma(T) = \mathbb{C} \setminus \rho(T)$, the spectrum $\sigma(T)$ is a closed set in $\mathbb{C}$. $\qquad\square$

Theorem 5.1.8. *Let H be a Hilbert space and let $T \in B(H)$ be a bounded linear operator on H. Then the spectrum $\sigma(T)$ for T is a compact set in $\mathbb{C}$, contained in the circle with radius $||T||$ and center 0.*

Proof. Assume that $|\lambda| > ||T||$. Then

$$T - \lambda I = -\lambda \left(I - \frac{1}{\lambda} T \right)$$

has a bounded inverse, since $||(1/\lambda)T|| < 1$, and hence $\lambda \in \rho(T)$. Consequently, $\lambda \in \sigma(T)$ implies that $|\lambda| \leq ||T||$, proving that $\sigma(T)$ is a bounded set in $\mathbb{C}$. But $\sigma(T)$ is also a closed set in $\mathbb{C}$, and hence $\sigma(T)$ is a compact set in $\mathbb{C}$. $\qquad\square$

Remark 5.1.9. For a bounded linear operator $T \in B(H)$ it follows in particular that $\lambda \in \rho(T)$ for $|\lambda| > ||T||$. Hence $\rho(T) \neq \emptyset$ for a bounded linear operator.

5.2 Spectral theorem for compact self-adjoint operators

We now turn our attention to bounded, self-adjoint linear operators T on a Hilbert space H. For such an operator T, we have the alternative formula for the operator norm $||T||$ proved in Theorem 4.1.11,

$$||T|| = \sup\{|(Tx, x)| \,|\, x \in H, ||x|| = 1\} .$$

Hence we get the following corollary from Theorem 5.1.8.

Corollary 5.2.1. *Let T be a bounded, self-adjoint linear operator on a Hilbert space H. Then for all $\lambda \in \sigma(T)$ in the spectrum for T it holds that*

$$|\lambda| \leq \sup\{|(Tx, x)| \,|\, x \in H, ||x|| = 1\} .$$

Example 5.2.2. Let (e_n) be an orthonormal basis for the infinite dimensional, separable Hilbert space H. Then the one-sided *shift operator* T on H with respect to this basis is defined by

$$T(x) = T\left(\sum_{n=1}^{\infty} (x, e_n) e_n \right) = \sum_{n=2}^{\infty} (x, e_n) e_{n-1} \quad \text{for all } x \in H .$$

We shall determine the spectrum $\sigma(T)$ for T. For that purpose, suppose

$$x = \sum_{n=1}^{\infty} (x, e_n) e_n \quad \text{is an eigenvector for } T ,$$

with corresponding eigenvalue λ. Then we get recursively,

$$(x, e_2) = \lambda(x, e_1)$$
$$(x, e_3) = \lambda(x, e_2) = \lambda^2(x, e_1)$$
$$(x, e_4) = \lambda(x, e_3) = \lambda^3(x, e_1)$$
$$\vdots$$
$$(x, e_n) = \lambda^{n-1}(x, e_1)$$
$$\vdots$$

Now further suppose that $(x, e_1) = 1$. Then we get $(x, e_n) = \lambda^{n-1}$ for all $n \in \mathbb{N}$, and hence $x = \sum_{n=1}^{\infty}(x, e_n)e_n$ is an eigenvector corresponding to $\lambda \in \mathbb{C}$ if and only if the sum defines a vector $x \in H$, which by Proposition 3.4.4 amounts to $\sum_{n=1}^{\infty}|\lambda^{n-1}|^2 < \infty$. The latter is certainly the case if and only if $|\lambda| < 1$. Hence the open unit disc is contained in the spectrum $\sigma(T)$.

On the other hand, for all $x \in H$, we have,

$$||Tx||^2 = \sum_{n=2}^{\infty}|(x, e_n)|^2 \leq \sum_{n=1}^{\infty}|(x, e_n)|^2 = ||x||^2 ,$$

and since $||Te_n|| = 1$ for $n \geq 2$, we get $||T|| = 1$.

By Theorem 5.1.8, it follows that the spectrum $\sigma(T)$ for the shift operator T must be contained in the closed unit disc. Since $\sigma(T)$ is a closed set in $\mathbb{C}$ containing the open unit disc, the spectrum $\sigma(T)$ for the shift operator T must be exactly the closed unit disc in $\mathbb{C}$.

The following theorem generalizes a well-known theorem from classical linear algebra in finite dimensions.

Theorem 5.2.3. *If T is a self-adjoint linear operator on the Hilbert space H, then all of its eigenvalues are real numbers. Furthermore, any pair of eigenvectors corresponding to different eigenvalues are orthogonal.*

Proof. Since T is self-adjoint (not necessarily bounded) we have,

$$(Tx, y) = (x, Ty) \quad \text{for all } x, y \in H .$$

If $Tx = \lambda x$ for $\lambda \in \mathbb{C}$ and a vector $x \neq 0$, we get

$$\lambda(x, x) = (\lambda x, x) = (Tx, x) = (x, Tx) = (x, \lambda x) = \overline{\lambda}(x, x) .$$

Since $(x, x) = ||x||^2 \neq 0$, we conclude that $\lambda = \overline{\lambda}$, proving that λ is real.

If now also $Ty = \mu y$ for $\mu \in \mathbb{C}$ and a vector $y \neq 0$, we get

$$\lambda(x, y) = (\lambda x, y) = (Tx, y) = (x, Ty) = (x, \mu y) = \overline{\mu}(x, y) = \mu(x, y) .$$

From this we conclude that $(x, y) = 0$ if $\lambda \neq \mu$.

This completes the proof of the theorem. $\qquad\qquad\qquad\qquad\square$

We are now heading for a proof that a self-adjoint, compact linear operator on a Hilbert space actually *has* an eigenvalue. We begin with the following result showing that at least asymptotically there is an eigenvalue.

Theorem 5.2.4. *Let T be a bounded, self-adjoint linear operator on a Hilbert space H. Then there exist a real number λ and a sequence (x_n) of unit vectors in H ($\|x_n\| = 1$ for all $n \in \mathbb{N}$), such that*

$$(T - \lambda I)x_n \to 0 \quad for \quad n \to \infty .$$

Furthermore, either $\lambda = \|T\|$ or $\lambda = -\|T\|$ has this property.

Proof. Since

$$\|T\| = \sup\{\,|(Tx, x)|\,\big|\, x \in H, \|x\| = 1\} ,$$

there is a sequence (y_n) with $\|y_n\| = 1$ in H such that

$$|(Ty_n, y_n)| \to \|T\| \quad \text{for} \quad n \to \infty .$$

Now, since (Ty_n, y_n) is a real number, there is a subsequence (x_n) of (y_n) for which either $(Tx_n, x_n) \to \|T\|$, or, $(Tx_n, x_n) \to -\|T\|$, for $n \to \infty$. Choose $\lambda = \pm\|T\|$ such that $(Tx_n, x_n) \to \lambda$ for $n \to \infty$.

Then we have

$$\begin{aligned}
\|(T - \lambda I)x_n\|^2 &= (Tx_n - \lambda x_n, Tx_n - \lambda x_n) \\
&= \|Tx_n\|^2 + \lambda^2\|x_n\|^2 - 2\lambda(Tx_n, x_n) \\
&\leq \|T\|^2 + \lambda^2 - 2\lambda(Tx_n, x_n) \\
&= 2\lambda^2 - 2\lambda(Tx_n, x_n) \to 2\lambda^2 - 2\lambda^2 = 0 ,
\end{aligned}$$

for $n \to \infty$. This proves that $\|(T - \lambda I)x_n\| \to 0$ for $n \to \infty$, and hence that $(T - \lambda I)x_n \to 0$ for $n \to \infty$. $\qquad\qquad\square$

Theorem 5.2.4 states that a bounded, self-adjoint linear operator T on a Hilbert space H has either $\|T\|$ or $-\|T\|$ as an approximate eigenvalue. In the limit, the sequence (x_n) functions as an eigenvector. The only problem is that the sequence (x_n) does not necessarily converge. When the operator T

is compact and self-adjoint, the sequence *does* converge and we get a proper eigenvalue as stated in the following theorem.

Theorem 5.2.5. *Let T be a compact, self-adjoint linear operator on a Hilbert space H. Then at least one of the numbers $\|T\|$ and $-\|T\|$ is an eigenvalue for T.*

Proof.   The case $T = 0$ is trivial, and hence we assume that $T \neq 0$.

Consider a sequence (x_n) in H chosen in accordance with Theorem 5.2.4 and related to $\lambda = \pm\|T\|$. Since T is compact, the sequence (x_n) has a subsequence (x_{n_k}) such that (Tx_{n_k}) is convergent. From the way (x_n) was chosen, it follows that $Tx_{n_k} - \lambda x_{n_k} \to 0$ for $k \to \infty$. Since $\lambda x_{n_k} = Tx_{n_k} - (Tx_{n_k} - \lambda x_{n_k})$ for all $k \in \mathbb{N}$, also the sequence (λx_{n_k}) must be convergent, and hence $x_{n_k} \to x_0$ for $k \to \infty$ for some $x_0 \in H$. By continuity of the norm $\|\cdot\|$, it follows that $\|x_0\| = 1$, since $\|x_{n_k}\| = 1$ for each $k \in \mathbb{N}$.

By the construction of $x_0 \in H$, it follows that $Tx_0 = \lambda x_0$. Since $x_0 \neq 0$, this proves that $\lambda = \pm\|T\|$ is an eigenvalue for T. $\square$

During the proof of Theorem 5.2.5, we have produced a unit vector $x_0 \in H$, such that $Tx_0 = \lambda x_0$, with $\lambda = \pm\|T\|$. From this we get,

$$\|Tx_0\| = \|\lambda x_0\| = |\lambda| \, \|x_0\| = \|T\|$$
$$= \sup\{|(Tx, x)| \,\big|\, x \in H, \|x\| = 1\},$$

showing that the supremum is attained for $x = x_0$ and hence is actually a maximum value. We get therefore the following corollary to Theorem 5.2.5.

Corollary 5.2.6. *Let T be a compact, self-adjoint linear operator on a Hilbert space H. Then the operator norm of T can be determined as a maximum value,*

$$\|T\| = \max\{|(Tx, x)| \,\big|\, x \in H, \|x\| = 1\}.$$

Moreover, the maximum value is attained for a nonzero eigenvector corresponding to the eigenvalue $\|T\|$ or $-\|T\|$.

Proposition 5.2.7. *Let T be a compact linear operator on the Hilbert space H, and assume that T has a nonzero eigenvalue λ. Then the corresponding eigenspace*

$$H_\lambda = \{x \in H \,|\, Tx = \lambda x\} \quad \text{is finite dimensional.}$$

Proof.   We proceed by an indirect proof. Assume therefore that H_λ is infinite dimensional. Then we can choose an orthonormal sequence (x_n) in H_λ, such that

$$\|Tx_n - Tx_m\|^2 = |\lambda|^2 \|x_n - x_m\|^2 = 2|\lambda|^2 > 0, \quad \text{for } n \neq m ,$$

proving that (Tx_n) contains no convergent subsequences. Hence T cannot be compact and we have a contradiction.

This proves that H_λ is finite dimensional. □

Then we are ready for the main theorem in this chapter.

Theorem 5.2.8 (Spectral Theorem - Compact Operators). *Let T be a compact, self-adjoint linear operator on a Hilbert space H of finite dimension, or of separable, infinite dimension. Then H admits an orthonormal basis (e_n) consisting of eigenvectors for T.*

In the finite dimensional case, the numbering of the finite sequence of basis vectors $(e_1, \ldots, e_k)$ can be chosen such that the corresponding finite sequence of eigenvalues $(\lambda_1, \ldots, \lambda_k)$ decreases numerically,

$$|\lambda_1| \geq |\lambda_2| \geq \ldots |\lambda_k| .$$

In the basis of eigenvectors, the operator T is described by,

$$Tx = \sum_{n=1}^{k} \lambda_n (x, e_n) e_n \quad \text{for} \quad x = \sum_{n=1}^{k} (x, e_n) e_n .$$

In the separable, infinite dimensional case, possible eigenvalues $\lambda_n = 0$ must be admitted anywhere in the sequence. With this tacitly understood, the numbering of the infinite sequence of basis vectors (e_n) can be chosen such that the sequence of corresponding eigenvalues (λ_n) decreases numerically,

$$|\lambda_1| \geq |\lambda_2| \geq \ldots |\lambda_n| \geq \ldots \quad \text{and} \quad \lambda_n \to 0 \text{ for } n \to \infty .$$

In the basis of eigenvectors, the operator T is described by,

$$Tx = \sum_{n=1}^{\infty} \lambda_n (x, e_n) e_n \quad \text{for} \quad x = \sum_{n=1}^{\infty} (x, e_n) e_n .$$

Proof. By Corollary 5.2.6, the compact, self-adjoint linear operator T admits at least one eigenvalue, namely

$$\lambda_1 = \pm \max\{ |(Tx, x)| \,\big|\, x \in H, \|x\| = 1 \} .$$

Choose a corresponding normalized eigenvector $e_1 \in H$.

Let $Q_1 = \{e_1\}^\perp$ be the orthogonal complement to the 1-dimensional subspace spanned by the vector e_1. Being an orthogonal complement, Q_1 is a closed subspace of the Hilbert space H and hence Q_1 is itself a Hilbert space. For $x \in Q_1$, we have the following computation

$$(Tx, e_1) = (x, Te_1) = \lambda_1(x, e_1) = 0 \ ,$$

showing that Q_1 is *invariant* under T in the sense that $Tx \in Q_1$ if $x \in Q_1$. Hence T can be considered as a compact, self-adjoint linear operator on the Hilbert space Q_1. Working inside Q_1, we then get a second eigenvalue for T,

$$\lambda_2 = \pm \max\{|(Tx, x)| \,|\, x \in Q_1, \|x\| = 1\} \ ,$$

and a normalized eigenvector $e_2 \in Q_1$ for λ_2. Clearly, $|\lambda_1| \geq |\lambda_2|$ and $e_1 \perp e_2$.

Next consider the orthogonal complement Q_2 to the 2-dimensional subspace spanned by the vectors e_1 and e_2. Working inside Q_2, we find a third eigenvalue λ_3 for T corresponding to a normalized eigenvector $e_3 \in Q_2$. Proceeding in this manner, we get an orthonormal sequence of eigenvectors (e_n) and a decreasing sequence,

$$|\lambda_1| \geq |\lambda_2| \geq \ldots |\lambda_n| \geq \ldots \ ,$$

defined by eigenvalues for T and associated with a sequence of subspaces

$$\cdots \subseteq Q_n \subseteq Q_{n-1} \subseteq \cdots \subseteq Q_1 \subseteq H,$$

such that each eigenvalue satisfies,

$$|\lambda_n| = \max\{|(Tx, x)| \,|\, x \in Q_{n-1}, \|x\| = 1\} \ .$$

If H is finite dimensional, say of dimension k, then this procedure terminates after k steps, and we have an orthonormal basis consisting of the k eigenvectors for T. If H is infinite dimensional, the orthonormal sequence (e_n) converges weakly to 0, and hence (Te_n) converges to 0 in norm since T is compact, thereby proving that $|\lambda_n| = \|Te_n\| \to 0$ for $n \to \infty$.

In the infinite dimensional case, let M be the linear subspace of H consisting of all convergent series $\sum_{n=1}^{\infty} \alpha_n e_n$. Then (e_n) is an orthonormal basis for M and via this basis, M can be identified with the complete space l^2. Hence M is a complete subspace, and therefore a closed subspace, in H. By the Projection Theorem 3.5.10, we can then decompose H as the orthogonal sum $H = M \oplus M^\perp$. Hence every vector $x \in H$ can be written uniquely in the form $x = z + y$, for vectors $z \in M$ and $y \in M^\perp$. Since every element $y \in M^\perp$ is orthogonal to all the eigenvectors e_n, the subspace $M^\perp$ of H is contained in all the subspaces Q_n of H.

Assertion 5.2.9. *Restricted to $M^\perp$, the operator T is the zero operator, i.e. $Ty = 0$ for all $y \in M^\perp$.*

Proof. The assertion is trivially true for $y = 0$. For an arbitrary nonzero vector $y \in M^\perp$ (if such a vector exists), we write y in the form $y = \|y\| y_1$, where $y_1 = y/\|y\|$. Then we get $(Ty, y) = \|y\|^2 (Ty_1, y_1)$. Since $y_1 \in Q_n$ for all $n \in \mathbb{N}$, it follows that $|(Ty_1, y_1)| \leq |\lambda_n|$ for all $n \in \mathbb{N}$, and consequently, that

$$|(Ty, y)| \leq \|y\|^2 |\lambda_n| \to 0 \quad \text{for} \quad n \to \infty ,$$

since $|\lambda_n| \to 0$ for $n \to \infty$. We conclude that $(Ty, y) = 0$ for all $y \in M^\perp$, from which follows (Exercise 56) that T is the zero operator on $M^\perp$. □

Now choose an orthonormal basis for $M^\perp$ supplementing the basis in M to an orthonomal basis for $H = M \oplus M^\perp$. Since the restriction of T to $M^\perp$ is the zero operator, each of the basis vectors from $M^\perp$ is an eigenvector for T corresponding to the eigenvalue 0. Finally, we assign numbers to the eigenvectors in the combined basis for H, for example by sticking the eigenvectors from $M^\perp$ in between the sequence of eigenvectors from M on the even numbers as long as needed to include all eigenvectors from $M^\perp$. In this way we obtain an orthonormal basis of eigenvectors for T with the properties in the theorem. In particular, we still have that $\lambda_n \to 0$ for $n \to \infty$, since all the new eigenvalues added to the sequence of eigenvalues from M are 0.

In the finite dimensional case, every vector $x \in H$ admits the (unique) expansion in the orthonormal basis of eigenvectors,

$$x = \sum_{n=1}^{k} (x, e_n) e_n .$$

Then we have

$$Tx = T\left(\sum_{n=1}^{k} (x, e_n) e_n \right) = \sum_{n=1}^{k} (x, e_n) Te_n = \sum_{n=1}^{k} \lambda_n (x, e_n) e_n .$$

In the separable, infinite dimensional case, every vector $x \in H$ admits the (unique) expansion in the orthonormal basis of eigenvectors,

$$x = \sum_{n=1}^{\infty} (x, e_n) e_n .$$

By continuity of T we then have

$$Tx = T\left(\lim_{k\to\infty} \sum_{n=1}^{k}(x,\,e_n)\,e_n\right) = \lim_{k\to\infty} T\left(\sum_{n=1}^{k}(x,\,e_n)\,e_n\right)$$

$$= \lim_{k\to\infty} \sum_{n=1}^{k}(x,\,e_n)\,Te_n = \sum_{n=1}^{\infty}\lambda_n\,(x,\,e_n)\,e_n \ .$$

This completes the proof of the Spectral Theorem. $\square$

Chapter 6

Fredholm Theory

In the late 19th century, the Italian Vito Volterra (1860–1940) and the Swede
Erik Ivar Fredholm (1866–1927) pushed the subject of integral equations forward in pioneering work that stimulated the development of functional analysis
as a theory of infinite dimensional function spaces and linear operators acting
on them. Fredholm introduced two types of integral equations that now carry
his name, and also lent his name to Fredholm operators.

In this chapter we present basic elements of the theory of Fredholm operators. We shall work within the framework of bounded operators on Banach
spaces, which is important for applications of Fredholm theory. This requires
that we develop the theory of dual spaces and dual operators between Banach
spaces to replace the use of adjoint operators between Hilbert spaces.

Fredholm operators are not only important in mathematical analysis, they
also play a significant role in theoretical physics, differential geometry and
topology with the famous Index Theorem proved by Michael Atiyah and
Isadore Singer in 1963 as a highlight. The magic of the Index Theorem lies
in the fact that it identifies an *analytical index* (related to the Fredholm index)
for an elliptic differential operator on a compact manifold with a *topological
index* of the manifold depending only on topological data of the manifold.

6.1 Dual spaces and operators

Throughout this chapter, X and Y is a pair of normed vector spaces over
$\mathbb{K} = \mathbb{R}, \mathbb{C}$, mostly Banach spaces or Hilbert spaces. If it is clear from the
context in which space we operate, we shall use the symbol $\|\cdot\|$ for a norm
and $(\cdot, \cdot)$ for an inner product without any label for the space.

For any bounded linear operator $T : X \to Y$ between Hilbert spaces X and
Y, we can define a unique bounded linear operator $T^* : Y \to X$, called the

adjoint operator of T, which satisfies the equation

$$(Tx, y)_Y = (x, T^*y)_X \quad \text{for all} \quad x \in X, y \in Y .$$

The adjoint operator shares all the properties of the adjoint operator introduced earlier for the case of a single Hilbert space H in Theorem 4.1.1, and all proofs go through verbatim. Also the useful Theorem 4.1.6 remains valid. We shall make use of this more general notion of an adjoint operator without further comments.

In the setting of Banach spaces there is a tool, which for many purposes can replace the use of adjoint operators between Hilbert spaces, namely the tool of dual operators.

The *dual space* of a normed vector space X is defined as the vector space X^* of all bounded linear functionals $\phi : X \to \mathbb{K}$ on X equipped with the operator norm

$$||\phi|| = \sup_{x \in X} \left\{ |\phi(x)| \mid ||x|| = 1 \right\} .$$

It is easy to prove that if X is a Banach space, then X^* with the operator norm is also a Banach space.

The *dual operator* of a linear operator $T : X \to Y$ between normed vector spaces is the linear operator $T^* : Y^* \to X^*$ defined by

$$(T^*\psi)(x) = \psi(Tx) \quad \text{for all} \quad \psi \in Y^*, \ x \in X .$$

It is easy to prove that if T is a bounded linear operator, then T^* is also a bounded linear operator.

Remark 6.1.1. If X and Y are Hilbert spaces, the dual operator $T^* : Y^* \to X^*$ of a bounded linear operator $T : X \to Y$ can be identified with the adjoint operator of T, using the Riesz' Representation Theorem 3.5.10.

There is a natural bilinear pairing

$$\langle \cdot, \cdot \rangle : X \times X^* \to \mathbb{K}$$

of a normed vector space X and its dual space X^*, defined by

$$\langle x, \phi \rangle = \phi(x) \quad \text{for all} \quad x \in X, \ \phi \in X^* .$$

The bilinear pairing $\langle \cdot, \cdot \rangle$ has many properties in common with an inner product, and satisfies among others an inequality like the Cauchy-Schwartz' inequality 3.1.11 for inner product spaces. Furthermore, the norms in the normed vector space and the dual space exhibit interesting dual expressions.

Lemma 6.1.2. *For the bilinear pairing $\langle \cdot, \cdot \rangle$ of a normed vector space X and its dual space X^* we have the inequality*

$$|\langle x, \phi \rangle| \leq ||x|| \, ||\phi|| \quad \text{for all} \quad x \in X, \; \phi \in X^* \, .$$

The norms in X and X^* have the dual expressions

$$||\phi|| = \sup_{x \in X} \left\{ |\phi(x)| \mid ||x|| = 1 \right\}$$

$$||x|| = \sup_{\phi \in X^*} \left\{ |\phi(x)| \mid ||\phi|| = 1 \right\} \, .$$

Proof. For arbitrary elements $x \in X$ and $\phi \in X^*$, we have

$$|\langle x, \phi \rangle| = |\phi(x)| \leq \sup_{x \in X} |\phi(x)| \leq ||\phi|| \, ||x|| \, ,$$

proving the inequality involving $\langle \cdot, \cdot \rangle$.

The given expression for $||\phi||$ is just the definition of the operator norm of ϕ. It therefore only remains to prove the expression for $||x||$.

Fix an arbitrary $x \neq 0 \in X$. Then we have

$$|\phi(x)| \leq ||\phi|| \, ||x|| \quad \text{for all} \quad \phi \in X^* \, ,$$

from which follows that

$$\sup_{\phi \in X^*} \left\{ |\phi(x)| \mid ||\phi|| = 1 \right\} \leq ||x|| \, .$$

Next, let U be the subspace of X spanned by x, and let $\lambda : U \to \mathbb{K}$ be the unique linear functional with $\lambda(x) = ||x||$. Clearly $||\lambda|| = 1$. By applying Corollary 1.9.2 we can extend λ to a linear functional $\rho : X \to \mathbb{K}$, with $||\rho|| = 1$ and $\rho(x) = ||x||$. Then

$$||x|| = \rho(x) \leq \sup_{\phi \in X^*} \left\{ |\phi(x)| \mid ||\phi|| = 1 \right\} \, ,$$

and we can conclude that

$$||x|| = \sup_{\phi \in X^*} \left\{ |\phi(x)| \mid ||\phi|| = 1 \right\} \, .$$

$\square$

For any bounded linear operator $T : X \to Y$ between normed vector spaces X and Y, the dual operator can by its very definition be characterized as the unique bounded linear operator $T^* : Y^* \to X^*$ satisfying the equation

$$\langle Tx, \psi \rangle = \langle x, T^*\psi \rangle \quad \text{for all} \quad x \in X, \psi \in Y^* \, .$$

Similar to an adjoint operator in Hilbert spaces, the operator norm of a dual operator in normed vector spaces also satisfies $||T^*|| = ||T||$.

Theorem 6.1.3. *Let $T : X \to Y$ be a bounded linear operator between normed vector spaces. Then the dual operator $T^* : Y^* \to X^*$ is also bounded and $||T^*|| = ||T||$.*

Proof.   From the definition of the dual operator, it follows that

$$||T^*\psi|| = \sup_{x \in X} \left\{ |\psi(Tx)| \mid ||x|| = 1 \right\} \le ||\psi|| \, ||T|| \,,$$

for all $\psi \in Y^*$, proving that T^* is bounded with $||T^*|| \le ||T||$.

By Corollary 1.9.2, we can for every $x_0 \in X$, choose $\psi_0 \in Y^*$ such that $||\psi_0|| = 1$ and $\psi_0(Tx_0) = ||Tx_0||$. Then

$$||Tx_0|| = \langle Tx_0 , \psi_0 \rangle = \langle x_0 , T^*\psi_0 \rangle \le ||T^*|| \, ||\psi_0|| \, ||x_0|| = ||T^*|| \, ||x_0|| \,,$$

proving that $||T|| \le ||T^*||$.

In conclusion we get $||T^*|| = ||T||$. $\square$

The following properties of dual operators are useful in many computations.

Theorem 6.1.4. *Let X, Y, Z be normed vector spaces.*

The dual operator of the composition $ST : X \to Z$ of bounded linear operators $T : X \to Y$ and $S : Y \to Z$ satisfies $(ST)^* = T^*S^*$.

The dual operator of the identity operator I_X on the normed vector space X satisfies $(I_X)^* = I_{X^*}$.

Proof.   The computation

$$\langle x , (ST)^*\psi \rangle = \langle STx , \psi \rangle = \langle Tx , S^*\psi \rangle = \langle x , T^*S^*\psi \rangle \,,$$

for $x \in X$ and $\psi \in Z^*$, shows that $(ST)^* = T^*S^*$.

The computation

$$\langle x , \phi \rangle = \langle (I_X)x , \phi \rangle = \langle x , (I_X)^*\phi \rangle \,,$$

for $x \in X$ and $\phi \in X^*$, shows that $(I_X)^* = I_{X^*}$. $\square$

Many results formulated (or proved) using adjoint operators in the setting of Hilbert spaces, remain valid in the setting of Banach spaces with almost identical proofs using dual operators.

In this connection double duals of Banach spaces and double duals of bounded linear operators play an important role. The constructions proceed as follows. For a bounded linear operator $T : X \to Y$, we can first construct the dual operator $T^* : Y^* \to X^*$, and then by repeating the process, the double dual operator $(T^*)^* : (X^*)^* \to (Y^*)^*$.

The norm in X^* is given by

$$\|\phi\| = \sup_{x \in X} \{|\phi(x)| \mid \|x\| = 1\} \quad \text{for all} \quad \phi \in X^* .$$

The norm in $(X^*)^*$ is given by

$$\|\xi\| = \sup_{\phi \in X^*} \{|\xi(\phi)| \mid \|\phi\| = 1\} \quad \text{for all} \quad \xi \in (X^*)^* .$$

There is a canonical map $J : X \to (X^*)^*$ defined by

$$\langle \phi, J(x) \rangle_{X^*} = \langle x, \phi \rangle_X \quad \text{for} \quad x \in X, \phi \in X^* .$$

The canonical map establishes a fruitful link from a Banach space to its double dual as described in the following theorem.

Theorem 6.1.5. *For any Banach space X, the map $J : X \to (X^*)^*$ is an injective, isometric, bounded linear operator, which maps X isomorphically onto the closed subspace $J(X)$ of the Banach space $(X^*)^*$.*

For every bounded linear operator $T : X \to Y$ between Banach spaces, the operator $(T^*)^* : (X^*)^* \to (Y^*)^*$ maps the closed subspace $J(X)$ of $(X^*)^*$ into the closed subspace $J(Y)$ of $(Y^*)^*$ such that $(T^*)^* J(x) = J(Tx)$ for all $x \in X$.

Proof. It is easy to prove that $J : X \to (X^*)^*$ is a linear operator. If $J(x) = 0$ for an element $x \in X$, then $\phi(x) = 0$ for all linear functionals $\phi \in X^*$, and hence $x = 0$ by Corollary 1.9.2, proving that J is injective.

The following computation proves that $J : X \to (X^*)^*$ is isometric and hence also a bounded linear operator. The computation use Lemma 6.1.2 in the final step:

$$\|J(x)\| = \sup_{\phi \in X^*} \{|J(x)(\phi)| \mid \|\phi\| = 1\}$$
$$= \sup_{\phi \in X^*} \{|\phi(x)| \mid \|\phi\| = 1\} = \|x\| .$$

Since $J : X \to (X^*)^*$ is an injective, isometric, bounded linear operator, it maps the Banach space X linearly isomorphic onto a complete, and hence closed, subspace $J(X)$ of the Banach space $(X^*)^*$. Then $J(X)$ is itself a Banach space, and the isomorphism is therefore not only a linear equivalence but also a topological equivalence (homeomorphism) by Banach's Theorem 1.8.3.

Now let $T : X \to Y$ be a bounded linear operator between Banach spaces. Then for all $x \in X$ and $\psi \in Y^*$ we have the computation

$$
\begin{aligned}
\langle J(x), (T^*)^* \psi \rangle_{X^*} &= \langle T^* \psi, J(x) \rangle_{X^*} \\
&= \langle x, T^* \psi \rangle_X \\
&= \langle Tx, \psi \rangle_Y = \langle \psi, J(Tx) \rangle_{Y^*} ,
\end{aligned}
$$

which shows that the operator $(T^*)^* : (X^*)^* \to (Y^*)^*$ maps the closed subspace $J(X)$ of $(X^*)^*$ into the closed subspace $J(Y)$ of $(Y^*)^*$ such that $(T^*)^* J(x) = J(Tx)$ for all $x \in X$.

This completes the proof of the theorem. $\qquad\square$

Compact operators play a key role in the study of operator equations. In this connection it is useful to know that the property of being compact is common for an operator and its dual operator.

Theorem 6.1.6. *Let $T : X \to Y$ be a bounded linear operator between Banach spaces, and let $T^* : Y^* \to X^*$ be the dual operator. Then T is a compact operator if and only if T^* is a compact operator.*

Proof.   First assume that T is a compact operator.

Let (ψ_n) be an arbitrary sequence in the open unit ball in Y^*, i.e. $\|\psi_n\| < 1$ for all $n \in \mathbb{N}$. In order to prove that T^* is a compact operator, it suffices to prove that the sequence $(T^*(\psi_n))$ contains a convergent subsequence with limit point in X^*.

Define the linear functional $\phi_n : X \to \mathbb{K}$ by $\phi_n(x) = \langle x, T^*(\psi_n) \rangle$ for all $x \in X$. By Lemma 6.1.2 and Theorem 6.1.3, we get

$$ |\phi_n(x) - \phi_n(y)| \leq \|T^*(\psi_n)(x - y)\| \leq \|T\| \|x - y\| , $$

proving that the sequence of functionals (ϕ_n) is equicontinuous.

Since T is a compact operator, the image of the open unit ball $B \subset X$ by T has compact closure $\overline{T(B)}$ in Y. By the Ascoli-Arzela Theorem 1.3.14, the sequence of continuous linear functionals (ϕ_n) contains therefore a subsequence (ϕ_{n_i}) that converges uniformly to a continuous function on $\overline{B}$. Since $\langle Tx, \psi_{n_i} \rangle = \langle x, T^*(\psi_{n_i}) \rangle$, this implies that the sequence of continuous linear functionals on X formed by the elements $T^*(\psi_{n_i}) \in X^*$ converges in norm to a continuous linear functional $\phi \in X^*$. This proves that the sequence $(T^*(\psi_{n_i}))$ is convergent with the limit point $\phi \in X^*$. Hence T^* is a compact operator.

Next assume that T^* is a compact operator. Then from the first part of the proof, $(T^*)^*$ must be compact. By Theorem 6.1.5, the operator T can be isometrically and linearly identified with the restriction of $(T^*)^*$ to the closed subspaces $J(X)$ of $(X^*)^*$ and $J(Y)$ of $(Y^*)^*$, which proves that T must also be a compact operator.

This completes the proof of the theorem. $\qquad\qquad\qquad\qquad\qquad\qquad$ $\square$

6.2 A duality theorem for operators with closed image

The main goal of this section is to prove a useful duality theorem for certain characteristic Banach spaces associated with a bounded linear operator (with closed image space) and its dual operator. As preparation for formulating and proving this duality theorem, we need to construct norms on quotient spaces of Banach spaces.

Recall that for an arbitrary subspace U of a vector space V, one can define a quotient vector space by identifying vectors $x, y \in V$ according to the equivalence relation

$$x \sim y \iff x - y \in U .$$

Since U is a subspace of V, it is easy to see that the vector space structure in V can be transferred to the set of equivalence classes defined by $\sim$. The vector space obtained by this construction is called the *quotient space* of V *modulo* U, and is denoted by $[V/U]$. The elements in $[V/U]$ are by construction equivalence classes $[x] = x + U$ of elements $x \in V$, and $[x] = [y]$ if and only if $x - y \in U$.

The surjective mapping $q : V \to [V/U]$, which maps a vector $x \in V$ onto its equivalence class $[x] \in [V/U]$, is called the *quotient mapping*. The quotient mapping is clearly linear in the vector space structure imposed on $[V/U]$.

In order to construct a positive definite norm for the quotient space $[V/U]$ induced by a norm in the vector space V, the subspace U has to be closed.

Lemma 6.2.1. *Let V be a normed vector space with norm $||\cdot||$, and let U be a closed linear subspace in V.*

Then the quotient space $[V/U]$ has an induced norm given by

$$||[x]|| = \inf_{u \in U} \{||x - u||\} \ for \ x \in V.$$

With this norm on $[V/U]$, the quotient mapping $q : V \to [V/U]$ is a surjective, bounded linear mapping with $||q(x)|| \le ||x||$ for all $x \in V$.

If V is a Banach space, then the quotient space $[V/U]$ with the induced norm is also a Banach space.

Proof. The quantity $||[x]|| = \inf_{u \in U}\{||x - u||\}$, is the distance of $x \in V$ to the closed subspace U. To prove that $||[\cdot]||$ defines a norm in $[V/U]$, we need to prove that it has the three properties NORM 1, NORM 2, NORM 3.

NORM 1. It is clear that $||\,[x]\,|| \geq 0$, so it only remains to prove that $||\,[x]\,|| = 0$ implies that $[x] = 0$, i.e. $x \in U$. This follows since $||\,[x]\,|| = 0$ implies that there exists a sequence (u_n) in U for which $||\,x - u_n\,|| \to 0$ for $n \to \infty$, which on the other hand implies that $x \in U$, since U is closed.

NORM 2. Follows since $\alpha U = U$ for every $\alpha \in \mathbb{K}$ with $\alpha \neq 0$.

NORM 3. The triangle inequality in $[V/U]$ follows from the triangle inequality in V by observing that for all $u, v \in U$ we have

$$
\begin{aligned}
||\,[x] + [y]\,|| = ||\,[x + y]\,|| \\
\leq ||\,x + y - u - v\,|| \\
\leq ||\,x - u\,|| + ||\,y - v\,|| ,
\end{aligned}
$$

which by definition of infimum implies that

$$||\,[x] + [y]\,|| - ||\,x - u\,|| \leq ||\,[y]\,|| ,$$

for all $u \in U$. And then finally again by definition of infimum that

$$||\,[x] + [y]\,|| \leq ||\,[x]\,|| + ||\,[y]\,|| ,$$

which is the triangle inequality in $[V/U]$.

With the induced norm on $[V/U]$, the quotient mapping $q : V \to [V/U]$ clearly satisfies $||\,q(x)\,|| \leq ||\,x\,||$ for all $x \in V$, and it is therefore a bounded, surjective linear mapping.

Now suppose that V is a Banach space. In order to prove that $[V/U]$ is a Banach space with the induced norm we have to prove that an arbitrary Cauchy sequence in $[V/U]$ is convergent.

Let $([x_n])$ be a Cauchy sequence in $[V/U]$. Using the Cauchy property, we can first by induction select a subsequence $([x_{n_k}])$ in $([x_n])$ such that $||\,[x_{n_{k+1}}] - [x_{n_k}]\,|| \leq 2^{-k-1}$ for all $k \geq 1$. Next we can, by definition of the induced norm in $[V/U]$, for all $k \geq 1$ choose $y_k \in [x_{n_{k+1}} - x_{n_k}]$ such that

$$||\,y_k\,|| \leq ||\,[x_{n_{k+1}}] - [x_{n_k}]\,|| + 2^{-k-1} \leq 2^{-k} .$$

Let $s_m = x_{n_1} + y_1 + \cdots + y_m$ be a partial sum in the series $x_{n_1} + \sum_{k=1}^{\infty} y_k$, which is convergent in the Banach space V, since it is majorized by the convergent series $||\,x_{n_1}\,|| + \sum_{k=1}^{\infty} 2^{-k}$ of real numbers. Hence the sequence of partial sums (s_m) converges to an element $x \in V$. We shall prove that the sequence $([x_n])$ is convergent with limit point $[x] \in [V/U]$.

From the relations $x_{n_1} \in [x_{n_1}]$ and $y_k \in [x_{n_{k+1}} - x_{n_k}]$, it follows by calculations with equivalence classes that $s_m \in [x_{n_{m+1}}]$, and therefore that

$$|| [x] - [x_{n_{m+1}}] || = || [x] - [s_m] || \leq || x - s_m || \ .$$

For all choices of elements x_n and $x_{n_{m+1}}$ above, we also have the inequality

$$|| [x] - [x_n] || \leq || [x] - [x_{n_{m+1}}] || + || [x_{n_{m+1}}] - [x_n] || \ .$$

Using the Cauchy property of the sequence $([x_n])$ and that $s_m \to x$ for $n \to \infty$, it follows from these inequalities, that for any given $\varepsilon > 0$ we can find $n_0 \in \mathbb{N}$ such that $|| [x] - [x_n] || < \varepsilon$ for all $n \geq n_0$. This proves that $([x_n])$ is convergent with limit point $[x] \in [V/U]$.

This completes the proof of the Lemma. $\square$

Before proceeding, we remind the reader about some terminology in connection with vector spaces associated with a linear operator $T : X \to Y$:

1. $\ker(T)$ is the *kernel* of T, i.e. the preimage $T^{-1}(\{0\})$ of $\{0\} \in Y$.
2. $T(X)$ is the *image* of $T : X \to Y$.
3. $\operatorname{coker}(T)$ is the *cokernel* of T, i.e. the quotient vector space $[Y/T(X)]$.

We also recall the definition of a direct sum of vector spaces.

Definition 6.2.2. Let V_1 and V_2 be linear subspaces of the vector space V. We say that V is the *direct sum* of V_1 and V_2, and write

$$V = V_1 \oplus V_2, \quad \text{if} \quad V = V_1 + V_2 \quad \text{and} \quad V_1 \cap V_2 = \{0\} \ .$$

Equivalently, $V = V_1 \oplus V_2$ if and only if every vector $v \in V$ admits a unique decomposition $v = v_1 + v_2$ for vectors $v_1 \in V_1$, $v_2 \in V_2$.

We also say that the subspaces V_1 and V_2 *complement* each other in the direct sum decomposition $V = V_1 \oplus V_2$.

It is often important to know whether a subspace admits a closed complement or not. As the following proposition shows, a finite dimensional subspace of a normed vector space always admits a closed complement.

Proposition 6.2.3. *Every finite dimensional subspace U of a normed vector space V with norm $||\cdot||$ can be complemented by a closed subspace $(V|U)$ of V in a direct sum decomposition*

$$V = (V|U) \oplus U \ .$$

The closed subspace $(V|U)$ of V can be identified linearly and topologically with the quotient space $[V/U]$ with the induced norm $\|[\cdot]\|$. In fact, the linear mapping, which maps $x \in (V|U)$ to $[x] \in [V/U]$, defines an equivalence *(isomorphism and homeomorphism) of normed vector spaces $(V|U) \equiv [V/U]$.*

Proof. Since U is a finite dimensional subspace of V, it is complete in the norm from V, and therefore a closed subspace of V.

Choose a finite basis $\{e_1, \ldots, e_n\}$ for U. We can define a dual basis for U^* consisting of n bounded linear functionals $\{\phi_1, \ldots, \phi_n\}$ on U, which satisfies

$$\langle e_j, \phi_i \rangle = \begin{cases} 0 \text{ for } i \neq j \\ 1 \text{ for } i = j. \end{cases}$$

By the Hahn-Banach Theorem 1.9.1, we can extend the bounded linear functionals $\{\phi_1, \ldots, \phi_n\}$ to V.

Define the bounded linear mapping $P : V \to V$ by

$$P(x) = \sum_{i=1}^{n} \langle x, \phi_i \rangle e_i \quad \text{for} \quad x \in V .$$

Clearly $P(V) = U$. By duality of the basis vectors, it follows that $P(x) = x$ for $x \in U$, and that P is idempotent, i.e. $P^2 = P$.

Every vector $x \in V$ can be written in the form

$$x = (I_V - P)(x) + P(x) ,$$

where I_V denotes the identity operator in V. Using that $P^2 = P$, it is now easy to prove that

$$V = \ker(P) \oplus U .$$

Let $(V|U) = \ker(P)$. Then $(V|U)$ is a closed subspace of V and we have

$$V = (V|U) \oplus U .$$

The linear mapping $q : (V|U) \to [V/U]$ defined by $q(x) = [x]$ for $x \in (V|U)$, is a bijective, bounded linear mapping between Banach spaces, and hence by Banach's Theorem 1.8.3, the inverse mapping q^{-1} is also bounded. This proves that q defines an equivalence (isomorphism and homeomorphism) of normed vector spaces $(V|U) \equiv [V/U]$.

This completes the proof. $\qquad\square$

In infinite dimensional normed vector spaces, linear subspaces are not necessarily closed, not even image spaces of bounded linear operators. The following proposition is therefore useful.

Proposition 6.2.4. *Let $T : X \to Y$ be a bounded linear operator between Banach spaces for which* $\operatorname{coker}(T)$ *is finite dimensional.*
Then the image space $T(X)$ of $T : X \to Y$ is a closed, linear subspace of Y, which can be complemented in Y by a closed subspace $(Y|T(X)) \equiv [Y/T(X)]$ in a direct sum decomposition

$$Y = T(X) \oplus (Y|T(X)) \ .$$

Proof. Since $\operatorname{coker}(T) = [Y/T(X)]$ is finite dimensional, Y admits algebraically a direct sum decomposition $Y = T(X) \oplus C$ for a finite dimensional, and hence closed, linear subspace C in Y. Note that C is a Banach space in the norm induced from Y. The quotient vector space $[X/\ker(T)]$ can be turned into a Banach space with norm $||[x]|| = \inf\{||x - x_0|| \, | \, x_0 \in \ker(T)\}$ by Lemma 6.2.1. Note that T induces an injective, bounded, linear operator $\tilde{T} : [X/\ker(T)] \to Y$. Now, define the linear mapping $S : [X/\ker(T)] \otimes C \to Y$ by $S([x], c) = T(x) + c$. By construction, S is a bijective, bounded linear mapping of the product Banach space $[X/\ker(T)] \otimes C$ onto the Banach space Y. Hence by Banach's Theorem 1.8.3, the linear mapping S is a topological equivalence (a homeomorphism) and therefore maps closed sets into closed sets. This proves that $T(X) = S([X/\ker(T)] \otimes \{0\})$ is a closed, linear subspace of Y.

The restriction of the quotient mapping $q : Y \to [Y/T(X)]$ to the subspace C in Y, is a bijective, bounded linear mapping between Banach spaces and hence defines an equivalence (isomorphism and homeomorphism) $C \equiv [Y/T(X)]$. Therefore $C = (Y|T(X))$ is a closed subspace in Y that complements $T(X)$ as described. □

We are now ready to prove a fundamental duality between the cokernel for a bounded operator with closed image and the kernel of its dual operator.

Theorem 6.2.5. *Let $T : X \to Y$ be a bounded linear operator between Banach spaces for which the image $T(X)$ is a closed, linear subspace of Y. Then there is an equivalence (isomorphism and homeomorphism) of Banach spaces*

$$\ker(T^*) \equiv \operatorname{coker}(T)^* \ .$$

Proof. The elements in the quotient vector space $\operatorname{coker}(T) = [Y/T(X)]$ are equivalence classes $[y]$ of elements in Y modulo $T(X)$. Since $T(X)$ is closed in Y, we can according to Lemma 6.2.1 equip $[Y/T(X)]$ with the norm $||[y]|| = \inf\{||y - T(x)|| \, | \, x \in X\}$ induced from the norm in Y. With this norm, $\operatorname{coker}(T)$ is a Banach space.

Every functional $\rho \in \ker(T^*) \subseteq Y^*$ induces a functional $\lambda \in \operatorname{coker}(T)^*$ by setting $\lambda([y]) = \rho(y)$ for each equivalence class $[y] \in \operatorname{coker}(T)$. The functional

λ is well defined since $\rho(Tx) = T^*(\rho)(x) = 0$ for all $x \in X$.

These considerations show that we get a well-defined mapping

$$\Phi : \ker(T^*) \to \operatorname{coker}(T)^*$$

by sending $\rho \in \ker(T^*)$ to the linear functional $\lambda \in \operatorname{coker}(T)^*$. It is easy to prove that Φ is a bounded linear mapping between Banach spaces.

We shall now define a mapping

$$\Psi : \operatorname{coker}(T)^* \to \ker(T^*)$$

which turns out to be inverse to Φ. The definition goes as follows.

Given a linear functional $\lambda : [Y/T(X)] \to \mathbb{K}$ in $\operatorname{coker}(T)^*$, we can follow the quotient mapping $q : Y \to [Y/T(X)]$ by λ to get a bounded linear functional $\rho = \lambda \circ q : Y \to \mathbb{K}$. Since $(T^*\rho)(x) = \rho(T(x)) = \lambda([T(x)]) = 0$ for all $x \in X$, the functional $\rho \in \ker(T^*)$. It is easy to prove that Ψ is a bounded linear mapping between Banach spaces, and that Ψ is inverse to Φ.

This completes the proof.   □

6.3   Fredholm Operators

For a linear operator $T : X \to Y$ between finite dimensional vector spaces X and Y, we have the following classical dimension formula

$$\dim(\ker(T)) - \dim(\operatorname{coker}(T)) = \dim(X) - \dim(Y) .$$

In infinite dimensions, the right-hand side of this formula does not make any sense. But what about the left-hand side? For an important class of linear operators, which we are about to define, the left-hand side of the formula makes sense also in infinite dimensions. These operators were introduced by Fredholm and now carry his name.

Definition 6.3.1. A bounded, linear operator $T : X \to Y$ between Banach spaces is said to be a *Fredholm operator* if it satisfies the conditions:

(1) The kernel of T is a finite dimensional linear subspace of X.
(2) The cokernel of T is a finite dimensional quotient vector space of Y.

If $T : X \to Y$ is a Fredholm operator, the *index* of T is the integer

$$\operatorname{index}(T) = \dim(\ker(T)) - \dim(\operatorname{coker}(T)) .$$

Remark 6.3.2. It is often included in the definition of a Fredholm operator $T : X \to Y$ that the image $T(X)$ of the operator shall be a closed subspace of Y. According to Proposition 6.2.4 this is redundant since it follows from the finite dimensionality of $\mathrm{coker}\,(T)$.

Example 6.3.3. Let $T : X \to X$ be an equivalence (isomorphism and homeomorphism) in an arbitrary normed vector space X, e.g. the identity operator I in X. Then $\ker\,(T) = \mathrm{coker}\,(T) = \{0\}$ showing that T is a Fredholm operator with $\mathrm{index}\,(T) = 0$.

Typical examples of Fredholm operators occur as perturbations of identity operators by compact operators.

Theorem 6.3.4. *Let X be an arbitrary Banach space, and let K be a compact linear operator on X. Then the bounded linear operator $T = I - K$ on X is a Fredholm operator with* $\mathrm{index}\,(T) = 0$.

Proof. By direct computation one checks that $K = I - T$ maps the closed unit ball in the closed subspace $\ker\,(T)$ of X identically onto itself. Since K is a compact operator, this proves that the closed unit ball in $\ker\,(T)$ is compact and hence $\ker\,(T)$ is finite dimensional by Theorem 4.2.3.

By Theorem 6.1.6, the dual operator K^* is also a compact operator, and hence we can argue with the dual operator $T^* = I_{X^*} - K^*$ exactly as with T and conclude that $\ker\,(T^*)$ is finite dimensional.

To prepare the way for making use of Theorem 6.2.5, we shall now prove that the image $T(X)$ is a closed linear subspace of X.

Consider the quotient vector space $[X/\ker\,(T)]$, which is a Banach space with the norm $\|\,[x]\,\| = \inf\{\|\,x - x_0\,\| \,|\, x_0 \in \ker\,(T)\}$, introduced in Lemma 6.2.1. The bounded linear operator T induces a bijective, bounded linear operator $\tilde{T} : [X/\ker\,(T)] \to T(X)$. The inverse linear mapping $\tilde{T}^{-1}$ is also a bounded linear operator, or equivalently a linear mapping continuous at $0 = T(0)$. We prove this indirectly. Suppose $\tilde{T}^{-1}$ is not continuous at $0 = T(0)$. Then there exist an $\varepsilon_0 > 0$ and a sequence (x_n) in X such that $\|T(x_n)\| < 1/n$ and $0 < \varepsilon_0 \leq \|\,[x_n]\,\| \leq \|x_n\| \leq 2\varepsilon_0$ for all $n \in \mathbb{N}$. Since the sequence (x_n) is bounded and K is a compact operator, we can assume by passage to a subsequence, that the sequence $(K(x_n))$ converges to an element $x \in X$. Since the sequence $(T(x_n))$ converges to 0 and $x_n = T(x_n) + K(x_n)$ for all $n \in \mathbb{N}$, the sequence (x_n) converges to the element $x \in X$. By continuity of T, the sequence $(T(x_n))$ converges to $T(x)$, and by construction to 0. Hence $T(x) = 0$, showing that $x \in \ker\,(T)$. But then $[x] = 0$. This contradicts that $\|\,[x_n]\,\| \geq \varepsilon_0 > 0$ and that $\|\,[x_n]\,\| \to \|\,[x]\,\| = 0$ for $n \to \infty$. Consequently, the inverse linear mapping $\tilde{T}^{-1}$ is continuous at $0 = T(0)$, and hence a bounded linear operator. Therefore $\tilde{T}$ is an equivalence of normed vector spaces, and since $[X/\ker(T)]$ is complete, the image $T(X)$ of $\tilde{T}$ must also be complete and therefore a closed linear subspace of X.

Since $T(X)$ is closed in Y and $\ker(T^*)$ is finite dimensional, $\operatorname{coker}(T)$ is finite dimensional by Theorem 6.2.5. This completes the proof that T is a Fredholm operator.

Then to the index of $T = I - K$. By Neumann's Lemma 5.1.6, it follows that if the compact operator K has operator norm $\|K\| < 1$, then $T = I - K$ has a bounded inverse and is an equivalence of normed vector spaces. Therefore $\ker(T) = \operatorname{coker}(T) = \{0\}$ and $\operatorname{index}(T) = 0$ for $\|K\| < 1$. For K an arbitrary compact operator, consider the operator $T_t = I - tK$, where t is a real number in the interval $0 \le t \le 1$. From what we have already proved, we know that $\{T_t\}$ is a parametrized family of Fredholm operators. Since $\|tK\| = t\|K\|$ we have $\operatorname{index}(T_t) = 0$ for $0 \le t\|K\| < 1$. Making appropriate use of Neumann's Lemma 5.1.6, it will be proved in Theorem 6.3.8 that the index function is locally constant and depends continuously on t. Therefore all the operators $\{T_t\}$ have the same index, and hence $\operatorname{index}(T) = \operatorname{index}(T_1) = \operatorname{index}(T_0) = 0$, as asserted in the theorem.   □

Fredholm operators can also be defined as bounded linear operators that are invertible modulo compact operators. This is the content of the following theorem, in which I_X and I_Y denote the identity operators in X and Y.

Theorem 6.3.5. *Let $T : X \to Y$ be a bounded, linear operator between Banach spaces. Then $T : X \to Y$ is a Fredholm operator if and only if there is a bounded linear operator $S : Y \to X$ such that the linear operators*

$$ST - I_X \quad and \quad TS - I_Y$$

are compact operators.

Proof.   First suppose that $T : X \to Y$ is a Fredholm operator.

From Propositions 6.2.3 and 6.2.4 we get that we can write

$$X = (X|\ker(T)) \oplus \ker(T) \quad and \quad Y = T(X) \oplus (Y|T(X))$$

for closed subspaces $(X|\ker(T)) \subseteq X$ and $(Y|T(X)) \subseteq Y$. Since $(X|\ker(T))$ can be identified with $[X/\ker(T)]$, we gleen from the proof of Theorem 6.3.4 that T by restriction to $(X|\ker(T))$ induces a bijective, bounded linear operator $\tilde{T} : (X|\ker(T)) \to T(X)$ for which the inverse mapping $\tilde{T}^{-1}$ is also a bounded linear operator. This inverse can be extended to a bounded linear operator

$S : Y \to X$ using the direct sum decompositions of X and Y. The operator $ST - I_X$ maps $X = (X|\ker(T)) \oplus \ker(T)$ into $\{0\} \oplus \ker(T)$ and therefore has finite rank. Similarly, the operator $TS - I_Y$ maps $Y = T(X) \oplus (Y|T(X))$ into $\{0\} \oplus (Y|T(X))$ and has finite rank. The bounded linear operator $S : Y \to X$ therefore satisfies the requirements in the theorem, since finite rank operators are compact operators by Proposition 4.2.6.

Next suppose that there exists a bounded linear operator $S : Y \to X$ such that $ST - I_X$ and $TS - I_Y$ are compact operators.

By direct computation one checks that the bounded linear operator $ST - I_X$ maps the closed unit ball in the Banach space $\ker(T)$ isometrically onto itself. Since $ST - I_X$ is a compact operator, it follows that the closed unit ball in $\ker(T)$ is compact, and hence $\ker(T)$ is finite dimensional by Theorem 4.2.3.

It now only remains to prove that $\operatorname{coker}(T)$ is finite dimensional.

Consider the nested triple of subspaces $T(S(Y)) \subseteq T(X) \subseteq Y$. Since the operator $TS - I_Y$ is compact, it follows by Theorem 6.3.4, that TS is a Fredholm operator and that the quotient space $[Y/T(S(Y))]$ is finite dimensional. Hence $[T(X)/T(S(Y))]$ is finite dimensional and therefore a closed subspace in $[Y/T(S(Y))]$, which implies that $T(X)$ is a closed subspace in Y, since $T(X) = q^{-1}([T(X)/T(S(Y))])$ is the preimage of the closed set $[T(X)/T(S(Y))] \subseteq [Y/T(S(Y))]$ for the (continuous) quotient mapping $q : Y \to [Y/T(S(Y))]$. According to Theorem 6.2.5, we then have the isomorphism $\ker(T^*) \equiv \operatorname{coker}(T)^*$.

By Theorem 6.1.6 and Theorem 6.3.4, we get that $\ker((TS)^*) = \ker(S^*T^*)$ is finite dimensional. And since $\ker(T^*) \subseteq \ker(S^*T^*)$ it follows that $\ker(T^*)$ is finite dimensional. Hence $\operatorname{coker}(T)$ is finite dimensional. This completes the proof that $T : X \to Y$ is a Fredholm operator. $\qquad\square$

Fredholm operators are closely linked to compact operators and finite rank operators. In the following theorem we show how to write a Fredholm operator as an operator matrix involving finite rank operators.

Theorem 6.3.6. *Let* $T : X \to Y$ *be a bounded, linear operator between Banach spaces. Then* $T : X \to Y$ *is a Fredholm operator if and only if*

(1) The Banach spaces X *and* Y *admit direct sum decompositions*

$$X = X_1 \oplus X_2 \quad and \quad Y = Y_1 \oplus Y_2$$

into closed linear subspaces, where X_2 *and* Y_2 *are finite dimensional.*
(2) With respect to the decompositions $X = X_1 \oplus X_2$ *and* $Y = Y_1 \oplus Y_2$ *the*

operator T admits a matrix description

$$T = \begin{bmatrix} T_{11} & T_{12} \\ T_{21} & T_{22} \end{bmatrix} : \begin{bmatrix} X_1 \\ X_2 \end{bmatrix} \to \begin{bmatrix} Y_1 \\ Y_2 \end{bmatrix}$$

where $T_{ij} : X_j \to Y_i$, $i, j = 1, 2$, are bounded linear operators and

(a) T_{11} is bijective and has a bounded inverse operator
(b) T_{12} and T_{22} are finite rank operators.

In such a decomposition of a Fredholm operator T we have

$$\text{index}(T) = \dim(X_2) - \dim(Y_2) .$$

Proof.   First suppose that $T : X \to Y$ is a Fredholm operator. In the proof of Theorem 6.3.5 we noticed that we can write

$$X = (X|\ker(T)) \oplus \ker(T) \quad \text{and} \quad Y = T(X) \oplus (Y|T(X))$$

for closed subspaces $(X|\ker(T)) \subseteq X$ and $(Y|T(X)) \subseteq Y$, where $\ker(T)$ and $(Y|T(X)) \equiv \text{coker}(T)$ are finite dimensional. We also noticed that restricting T to $(X|\ker(T))$ induces a bijective, linear operator $\tilde{T} : (X|\ker(T)) \to T(X)$ for which the inverse operator is bounded.

Put $X_1 = (X|\ker(T))$, $X_2 = \ker(T)$, $Y_1 = T(X)$ and $Y_2 = (Y|T(X))$. Then the direct sum decompositions

$$X = X_1 \oplus X_2 \quad \text{and} \quad Y = Y_1 \oplus Y_2$$

satisfies the conditions in (1). By restricting T to X_j and composing with the projection of $Y = Y_1 \oplus Y_2$ onto Y_i, we get a matrix of bounded linear operators $T_{ij} : X_j \to Y_i$, $i, j = 1, 2$, which satisfies the conditions in (2).

Next suppose that $T : X \to Y$ admits a matrix description satisfying conditions (1) and (2).

Let $S : Y \to X$ be the bounded linear operator with the matrix description

$$S = \begin{bmatrix} T_{11}^{-1} & 0 \\ 0 & 0 \end{bmatrix}$$

with respect to the decompositions $X = X_1 \oplus X_2$ and $Y = Y_1 \oplus Y_2$. Then we have the following computations, in which I throughout denotes an identity operator in an appropriate subspace of X or Y:

$$ST = \begin{bmatrix} T_{11}^{-1} & 0 \\ 0 & 0 \end{bmatrix} \begin{bmatrix} T_{11} & T_{12} \\ T_{21} & T_{22} \end{bmatrix} = \begin{bmatrix} I & T_{11}^{-1}T_{12} \\ 0 & 0 \end{bmatrix} = \begin{bmatrix} I & 0 \\ 0 & I \end{bmatrix} + \begin{bmatrix} 0 & T_{11}^{-1}T_{12} \\ 0 & -I \end{bmatrix}$$

$$TS = \begin{bmatrix} T_{11} & T_{12} \\ T_{21} & T_{22} \end{bmatrix} \begin{bmatrix} T_{11}^{-1} & 0 \\ 0 & 0 \end{bmatrix} = \begin{bmatrix} I & 0 \\ T_{21}T_{11}^{-1} & 0 \end{bmatrix} = \begin{bmatrix} I & 0 \\ 0 & I \end{bmatrix} + \begin{bmatrix} 0 & 0 \\ T_{21}T_{11}^{-1} & -I \end{bmatrix}.$$

The computations show that the operators $ST - I$ and $TS - I$ have finite rank. By Proposition 4.2.6 finite rank operators are compact and hence it follows by Theorem 6.3.5 that T is a Fredholm operator.

Finally, we turn to determine the index of T.

Let $F : X_2 \to Y_2$ denote the operator $F = T_{22} - T_{21}T_{11}^{-1}T_{12}$. Simple computations show that

$$\ker(T) = \left\{ x_1 + x_2 \in X_1 \oplus X_2 \mid x_1 = -T_{11}^{-1}T_{12}(x_2), \ \ x_2 \in \ker(F) \right\}.$$

It is then easy to prove that the linear mapping $L = I_{X_2} - T_{11}^{-1}T_{12}$ defines an equivalence of normed vector spaces $\ker(F) \equiv \ker(T)$.

For the dual operator T^* of T, we have the matrix description

$$T = \begin{bmatrix} T_{11}^* & T_{21}^* \\ T_{12}^* & T_{22}^* \end{bmatrix} : \begin{bmatrix} Y_1^* \\ Y_2^* \end{bmatrix} \to \begin{bmatrix} X_1^* \\ X_2^* \end{bmatrix}$$

which satisfies conditions corresponding to conditions (1) and (2) for T. Substituting T by T^* we can therefore conclude that $\ker(F^*) \equiv \ker(T^*)$. By Theorem 6.2.5 this implies that $\operatorname{coker}(F) \equiv \operatorname{coker}(T)$.

Since $F : X_2 \to Y_2$ is a linear mapping between finite dimensional vector spaces, we get now by a classical dimension formula in finite dimensions that

$$\operatorname{index}(T) = \dim(\ker(F)) - \dim(\operatorname{coker}(F)) = \dim(X_2) - \dim(Y_2).$$

This completes the proof of the theorem. $\qquad\square$

We finish this section with proofs of some fundamental properties of Fredholm operators.

Proposition 6.3.7. *If $T : X \to Y$ is a Fredholm operator between Banach spaces X and Y, then the dual operator $T^* : Y^* \to X^*$ is also a Fredholm operator, and*

$$\operatorname{index}(T^*) = -\operatorname{index}(T).$$

Proof. In the proof of Theorem 6.3.6 we made use of the matrix description of the dual operator T^* corresponding to the matrix description of T, and remarked that it satisfies the conditions (1) and (2) for being the matrix of a Fredholm operator. Therefore T^* is a Fredholm operator. From the matrix description of T^* we can determine the index of T^* by the computation

$$\text{index}(T^*) = \dim(Y_2^*) - \dim(X_2^*) = \dim(Y_2) - \dim(X_2) = -\text{index}(T) . \quad \square$$

Proposition 6.3.8. *Let* $T : X \to Y$ *and* $K : X \to Y$ *be bounded linear operators between Banach spaces* X *and* Y.

(1) *If* T *is a Fredholm operator and if the operator norm* $\|K\|$ *of* K *is sufficiently small, then the operator* $T + K$ *is a Fredholm operator and* $\text{index}(T + K) = \text{index}(T)$.

(2) *If* T *is a Fredholm operator and if* K *is compact, then* $T + K : X \to Y$ *is a Fredholm operator, and* $\text{index}(T + K) = \text{index}(T)$.

Proof. **Assertion (1)** By Theorem 6.3.6 we know that T can be given a matrix description

$$T = \begin{bmatrix} T_{11} & T_{12} \\ T_{21} & T_{22} \end{bmatrix} : \begin{bmatrix} X_1 \\ X_2 \end{bmatrix} \to \begin{bmatrix} Y_1 \\ Y_2 \end{bmatrix} ,$$

with respect to direct sum decompositions $X = X_1 \oplus X_2$ and $Y = Y_1 \oplus Y_2$ of the Banach spaces X and Y into closed linear subspaces. Here X_2 and Y_2 are finite dimensional, and $T_{11} : X_1 \to Y_1$ is bijective with a bounded inverse operator.

With respect to the decompositions $X = X_1 \oplus X_2$ and $Y = Y_1 \oplus Y_2$, we can give K the matrix description

$$K = \begin{bmatrix} K_{11} & K_{12} \\ K_{21} & K_{22} \end{bmatrix} : \begin{bmatrix} X_1 \\ X_2 \end{bmatrix} \to \begin{bmatrix} Y_1 \\ Y_2 \end{bmatrix}$$

By choosing $\|K\|$ sufficiently small, Neumann's Lemma 5.1.6 ensures that $T_{11} + K_{11}$ is still bijective with bounded inverse. [In fact, $\|T_{11}^{-1}\| \|K\| < 1$ suffices; see proof of Theorem 6.4.4.] But then $T + K$ has a matrix description

$$T + K = \begin{bmatrix} T_{11} + K_{11} & T_{12} + K_{12} \\ T_{21} + K_{21} & T_{22} + K_{22} \end{bmatrix} : \begin{bmatrix} X_1 \\ X_2 \end{bmatrix} \to \begin{bmatrix} Y_1 \\ Y_2 \end{bmatrix} ,$$

which by Theorem 6.3.6 is characteristic for a Fredholm operator, and

$$\text{index}(T + K) = \dim(X_2) - \dim(Y_2) = \text{index}(T) .$$

Assertion (2) Making use of Theorem 6.3.5 it follows easily that $T + K$ is a Fredholm operator.

For the index, consider the family of Fredholm operators $T_t = T + tK$, for $0 \le t \le 1$. Arguing again with a matrix description of the Fredholm operator $T_{t_0} = T + t_0 K$ for a fixed, but arbitrarily chosen, $0 \le t_0 \le 1$, it follows that the index function is constant in a neighborhood of t_0. This proves that the index function is locally constant and depends continuously on t. Therefore all the operators $\{T_t\}$ have the same index, and

$$\text{index}(T + K) = \text{index}(T_1) = \text{index}(T_0) = \text{index}(T) .$$

This completes the proof of the theorem. $\qquad\qquad\qquad\qquad\qquad\qquad\square$

Proposition 6.3.9. *Let X, Y, Z be Banach spaces, and let $T : X \to Y$ and $S : Y \to Z$ be bounded linear operators.*

    (1) If at least one of the operators T and S is compact, then the composition $ST : X \to Z$ is a compact operator.

    (2) If both of the operators T and S are Fredholm operators, then the composition $ST : X \to Z$ is a Fredholm operator, and

$$\text{index}(ST) = \text{index}(T) + \text{index}(S) .$$

Proof. **Assertion (1)** is an immediate consequence of the fact that the image of a compact set under a continuous mapping is compact.

Assertion (2) Let $T : X \to Y$ and $S : Y \to Z$ be Fredholm operators. Using that kernels of operators are preimages of $\{0\}$, we have the computation

$$\ker(ST) = (ST)^{-1}(\{0\}) = T^{-1}(S^{-1}(\{0\})) = T^{-1}(\ker(S)) ,$$

showing that we have an identification of quotient spaces

$$[\ker(ST)/\ker(T)] \equiv [T^{-1}(\ker(S))/\ker(T)] .$$

Since T and S are Fredholm operators, we get that $[T^{-1}(\ker(S))/\ker(T)]$ is a finite dimensional vector space. From this follows that $\ker(ST)$ is finite dimensional, and that $\dim(\ker(ST)) \le \dim(\ker(T)) + \dim(\ker(S))$.

By Proposition 6.3.7, the dual operators T^* and S^* are also Fredholm operators, and therefore by similar arguments $\ker((ST)^*) = \ker(T^*S^*)$ is finite dimensional. Hence by Theorem 6.2.5, $\text{coker}(ST)$ is finite dimensional. This proves that the composition ST is a Fredholm operator.

To prove the formula for the index, we consider a family of linear operators $L_t : Y \oplus X \to Z \oplus Y$, parametrized by t in the interval $0 \le t \le 1$, and defined by the matrix desciption

$$L_t = \begin{bmatrix} \cos(\frac{\pi}{2}t)S & -\sin(\frac{\pi}{2}t)ST \\ \sin(\frac{\pi}{2}t)I & \cos(\frac{\pi}{2}t)T \end{bmatrix} : \begin{bmatrix} Y \\ X \end{bmatrix} \rightarrow \begin{bmatrix} Z \\ Y \end{bmatrix}.$$

The matrix description of L_t can be factored as follows

$$L_t = \begin{bmatrix} S & 0 \\ 0 & I \end{bmatrix} \begin{bmatrix} \cos(\frac{\pi}{2}t)I & -\sin(\frac{\pi}{2}t)I \\ \sin(\frac{\pi}{2}t)I & \cos(\frac{\pi}{2}t)I \end{bmatrix} \begin{bmatrix} I & 0 \\ 0 & T \end{bmatrix}.$$

Now that we know a composition of Fredholm operators is again a Fredholm operator, it is easy to prove that $\{L_t\}$ is a family of Fredholm operators. From the proof of Theorem 6.3.8 we know that the index function is locally constant and depends continuously on t. Hence all the operators in the family $\{L_t\}$ have the same index, and we get

$$\text{index}\,(T) + \text{index}\,(S) = \text{index}\,(L_0) = \text{index}\,(L_1) = \text{index}\,(ST)\,.$$

□

Example 6.3.10. Let H be an infinite dimensional, separable Hilbert space with orthonormal basis (e_n).

Define the unilateral shift $S : H \rightarrow H$ by $S(e_n) = e_{n+1}$ for all $n \in \mathbb{N}$. Clearly, S is a bounded linear operator with $\ker(S) = \{0\}$ and $\text{coker}(S)$ the 1-dimensional subspace spanned by e_1. Therefore S is a Fredholm operator with $\text{index}(S) = -1$. By Proposition 6.3.9, it follows that the n-fold iteration S^n of S is a Fredholm operator with $\text{index}(S^n) = -n$ for all $n \in \mathbb{N}$.

The dual operator S^* of S is the one-sided shift operator T defined in Example 5.2.2. By Proposition 6.3.7, we get that S^* is a Fredholm operator with $\text{index}(S^*) = 1$, and further by Proposition 6.3.9, that $(S^*)^n$ is a Fredholm operator with $\text{index}((S^*)^n) = n$ for all $n \in \mathbb{N}$.

6.4   Topology of spaces of operators

In this section we shall discuss topological properties of some of the spaces of operators we have introduced in the previous sections.

For Banach spaces X and Y and bounded linear operators $T : X \rightarrow Y$, we shall use the following notations for spaces of operators.

$\mathcal{B}(X,Y)$ is the space of all bounded operators.
$\mathcal{C}(X,Y)$ is the subset of compact operators.
$\mathcal{F}(X,Y)$ is the subset of Fredholm operators.
$\mathcal{I}(X,Y)$ is the subset of invertible operators.

If $X = Y$, we just write $\mathcal{B}(X)$, $\mathcal{C}(X)$, $\mathcal{F}(X)$ and $\mathcal{I}(X)$ for the spaces.

All subsets of a space of bounded operators have metric structure induced from the operator norm in $\mathcal{B}(X, Y)$.

Theorem 6.4.1. *For any pair of Banach spaces X and Y, the space of bounded operators $\mathcal{B}(X, Y)$ is a Banach space, when equipped with the operator norm.*

Proof. From Theorem 1.7.7 we know that $\mathcal{B}(X, Y)$ is a normed vector space, when equipped with the operator norm.

To prove that $\mathcal{B}(X, Y)$ is a Banach space, we need to prove that an arbitrarily given Cauchy sequence (T_n) of bounded operators is convergent in the operator norm. With this in mind, first note that for all $x \in X$

$$\|T_n(x) - T_m(x)\| = \|(T_n - T_m)(x)\| \le \|T_n - T_m\| \, \|x\| \ .$$

This shows that the sequence $(T_n(x))$ is a Cauchy sequence in Y for all $x \in X$. Since Y is a Banach space, the sequence $(T_n(x))$ is convergent in Y with a unique limit point, and hence we can define a mapping $T : X \to Y$ by

$$T(x) = \lim_{n \to \infty} T_n(x) \quad \text{for all} \quad x \in X \ .$$

Since all the operators T_n are linear it follows that the limit mapping T is also a linear operator. The inequality

$$\big| \|T_n\| - \|T_m\| \big| \le \|T_n - T_m\|$$

shows that the sequence of real numbers $(\|T_n\|)$ is a Cauchy sequence in $\mathbb{R}$, and hence convergent to $c \in \mathbb{R}$. From the inequality $\|T_n(x)\| \le \|T_n\| \, \|x\|$, we get by taking limits that $\|T(x)\| \le c\|x\|$, showing that T is bounded and that

$$\|T\| = \lim_{n \to \infty} \|T_n\| \ .$$

We shall now prove that (T_n) converges to T.

Let $x \in X$ be an arbitrary element in X with $\|x\| = 1$. For all $n, k \in \mathbb{N}$, we get by the triangle inequality

$$\|(T - T_n)(x)\| \le \|T(x) - T_{n+k}(x)\| + \|T_{n+k}(x) - T_n(x)\|$$
$$\le \|T(x) - T_{n+k}(x)\| + \|T_{n+k} - T_n\| \ .$$

Given $\varepsilon > 0$, choose $n_0 \in \mathbb{N}$ such that $\|T_{n+k} - T_n\| \le \varepsilon$ for $n \ge n_0$. Taking the limit $k \to \infty$ it follows that $\|(T - T_n)(x)\| \le \varepsilon$ for $n \ge n_0$. From this we conclude that $\|T - T_n\| \le \varepsilon$ for $n \ge n_0$, proving that (T_n) converges to T. □

We have earlier in the text used the word *equivalence* for a correspondence between normed vector spaces established by a bijective, bounded linear operator with a bounded inverse operator. Such operators are said to be invertible.

Definition 6.4.2. A bounded linear operator $T : X \to Y$ between normed vector spaces X and Y is said to be *invertible* if it is bijective and if it has a bounded inverse operator $T^{-1} : Y \to X$.

Remark 6.4.3. A bounded linear operator $T : X \to Y$ between Banach spaces X and Y is invertible if and only if it is bijective; see Banach's Theorem 1.8.3.

Theorem 6.4.4. *The set of invertible operators $\mathcal{I}(X, Y)$ is an open set in $\mathcal{B}(X, Y)$ for any pair of Banach spaces X and Y.*

Proof. If $\mathcal{I}(X, Y) = \emptyset$, the theorem is trivially true. Suppose therefore that $\mathcal{I}(X, Y) \neq \emptyset$, and let $T : X \to Y$ be an invertible operator.

For an arbitrary bounded operator $K : X \to Y$ we can write

$$T + K = T(T^{-1}T + T^{-1}K) = T(I_X + T^{-1}K) .$$

By Neumann's Lemma 5.1.6 it follows that $T + K$ is invertible if

$$\|T^{-1}K\| \leq \|T^{-1}\| \|K\| < 1 .$$

Hence the open ball with center T and radius $r > 0$ is contained in $\mathcal{I}(X, Y)$ if $0 < r\|T^{-1}\| < 1$, proving that $\mathcal{I}(X, Y)$ is an open set in $\mathcal{B}(X, Y)$. □

Remark 6.4.5. For a finite dimensional Banach space X over $\mathbb{K}$, the space of invertible operators $\mathcal{I}(X)$ is connected if $\mathbb{K} = \mathbb{C}$. For $\mathbb{K} = \mathbb{R}$ it has two connected components, classified by the sign of the determinant. If X is infinite dimensional, the French mathematician Adrian Douady gave in 1965 examples that $\mathcal{I}(X)$ may have infinitely many components. However, for a large class of Banach spaces, including infinite dimensional Hilbert spaces H, it was proved the same year by the Dutch mathematician N.H. Kuiper that $\mathcal{I}(H)$ is contractible to the identity, in particular that it is connected.

The following result about spaces of invertible operators is due to Putnam and Wintner [Proc. Nat. Acad. Sci., Wash. **37**(1951), 110-112], and was proved using spectral resolutions of operators. We state the theorem without proof.

Theorem 6.4.6. *The space of invertible operators $\mathcal{I}(H)$ on an infinite dimensional, separable Hilbert space H is connected.*

We now turn to spaces of compact operators.

Lemma 6.4.7. *A linear combination of compact operators between normed vector spaces is again a compact operator.*

Proof. Let X and Y be normed vector spaces. Consider a linear combination $\alpha S + \beta T$ of compact operators $S, T : X \to Y$, where $\alpha, \beta \in \mathbb{K}$. Let (x_n) be an arbitrary bounded sequence in X. Since S and T are compact, we can select a subsequence (x_{n_k}) of (x_n), such that both of the sequences $(S(x_{n_k}))$ and $(T(x_{n_k}))$ are convergent. But then the sequence $((\alpha S + \beta T)(x_{n_k}))$ is convergent, proving that $\alpha S + \beta T$ is a compact operator. □

Theorem 6.4.8. *The set of compact operators $\mathcal{C}(X, Y)$ is a closed linear subspace in $\mathcal{B}(X, Y)$ for any pair of Banach spaces X and Y.*

Proof. That $\mathcal{C}(X, Y)$ is a subspace of $\mathcal{B}(X, Y)$ follows by Lemma 6.4.7.

To prove that $\mathcal{C}(X, Y)$ is a closed subspace of $\mathcal{B}(X, Y)$ we proceed as in the proof of Theorem 4.2.11.

Let (T_n) be a sequence of operators in $\mathcal{C}(X, Y)$ which converges (in operator norm) to the operator $T \in \mathcal{B}(X, Y)$. To prove that $\mathcal{C}(X, Y)$ is closed, we just need to prove that T is a compact linear operator. This amounts to proving that if (x_n) is an arbitrary bounded sequence in X, then we can extract a subsequence (x_{n_k}) such that the sequence (Tx_{n_k}) is convergent in Y.

Let (x_n) be an arbitrary bounded sequence in X. Since T_1 is compact, we can extract a subsequence (x_n^1) of (x_n) such that $(T_1 x_n^1)$ is convergent. Since T_2 is compact, we can extract a subsequence (x_n^2) of (x_n^1) such that $(T_2 x_n^2)$ is convergent. By induction, we get for every $k \geq 2$ a subsequence (x_n^k) of (x_n^{k-1}) such that $(T_k x_n^k)$ is convergent. The diagonal sequence (x_n^n) is a subsequence of (x_n), and it is obvious that the sequence $(T_k x_n^n)$ is convergent for every $k \geq 1$.

If we can prove that the sequence (Tx_n^n) is convergent in Y, it follows that T is a compact operator. Since Y is a Banach space, it is sufficient to prove that (Tx_n^n) is a Cauchy sequence in Y.

With this in mind, let $\varepsilon > 0$ be given. Since the original sequence (x_n) is bounded, there exists a positive constant C such that $\|x_n^n\| \leq C$ for all $n \in \mathbb{N}$. Since $T_n \to T$ for $n \to \infty$ (in operator norm), we can choose a number $k_0 \in \mathbb{N}$, which we then keep fixed, such that

$$\|T - T_{k_0}\| \leq \frac{\varepsilon}{3C} .$$

Since $(T_{k_0} x_n^n)$ is convergent, there exists a number $n_0 \in \mathbb{N}$, such that

$$\|T_{k_0} x_n^n - T_{k_0} x_m^m\| \leq \frac{\varepsilon}{3} \quad \text{for all} \quad n, m \geq n_0 .$$

For $n, m \geq n_0$ we now have the computation

$$\|Tx_n^n - Tx_m^m\| \leq \|Tx_n^n - T_{k_0}x_n^n\| + \|T_{k_0}x_n^n - T_{k_0}x_m^m\| + \|T_{k_0}x_m^m - Tx_m^m\|$$

$$\leq \|T - T_{k_0}\|\|x_n^n\| + \frac{\varepsilon}{3} + \|T_{k_0} - T\|\|x_m^m\|$$

$$\leq \frac{\varepsilon}{3C}C + \frac{\varepsilon}{3} + \frac{\varepsilon}{3C}C = \varepsilon .$$

This proves that (Tx_n^n) is indeed a Cauchy sequence in Y, and hence that the sequence (Tx_n^n) is convergent in Y, proving that T is a compact linear operator.

This completes the proof of the theorem. $\qquad \square$

Theorem 6.4.9. *For any Banach space X, the set of compact operators $\mathcal{C}(X)$ is a closed, bilateral ideal in the Banach algebra $\mathcal{B}(X)$.*

Proof. $\mathcal{B}(X)$ is an algebra with the vector space structure defined by addition and scalar multiplication of bounded operators, and product defined by composition of bounded operators. $\mathcal{C}(X)$ is a closed, bilateral ideal in $\mathcal{B}(X)$, since by Theorem 6.4.8, it is a closed subspace in $\mathcal{B}(X)$, and since by Proposition 6.3.9, the product of a compact operator with an arbitrary bounded operator from either side is again a compact operator. $\qquad \square$

Corollary 6.4.10. *For any Banach space X, the quotient space $[\mathcal{B}(X)/\mathcal{C}(X)]$ is a Banach algebra, the Calkin algebra, and the quotient mapping*

$$q : \mathcal{B}(X) \to [\mathcal{B}(X)/\mathcal{C}(X)]$$

is a surjective, bounded Banach algebra homomorphism.

A bounded operator $T \in \mathcal{B}(X)$ is a Fredholm operator if and only if the element $q(T)$ is invertible in $[\mathcal{B}(X)/\mathcal{C}(X)]$.

Proof. The first part of the corollary is an immediate consequence of Lemma 6.2.1 and Theorem 6.4.9. The second part is a rephrasing of Theorem 6.3.5.$\square$

Finally, we turn the attention to spaces of Fredholm operators.

Theorem 6.4.11. *For any pair of Banach spaces X and Y, the set of Fredholm operators $\mathcal{F}(X,Y)$ is an open set in $\mathcal{B}(X,Y)$, and the function*

$$\text{index} : \mathcal{F}(X,Y) \to \mathbb{Z} ,$$

that assigns the index to every Fredholm operator, is locally constant and continuous, and therefore constant in every connected component of $\mathcal{F}(X,Y)$.

Proof. Let $T : X \to Y$ be a Fredholm operator. For an arbitrary bounded linear operator $K : X \to Y$, we know by Proposition 6.3.8 that if the operator norm of K is sufficiently small, then the operator $T+K$ is a Fredholm operator and index $(T+K) =$ index (T).

This shows that for a sufficiently small radius $r > 0$, the open ball with center T and radius $r > 0$ in $\mathcal{B}(X,Y)$ is contained in $\mathcal{F}(X,Y)$, proving that $\mathcal{F}(X,Y)$ is an open set in $\mathcal{B}(X,Y)$, and that the index function is locally constant and continuous in $\mathcal{F}(X,Y)$. The latter implies in particular, that the index function is constant in every connected component of $\mathcal{F}(X,Y)$. $\square$

Proposition 6.4.12. *For a bounded operator* $T : X \to Y$ *between Banach spaces* X *and* Y *the following statements are equivalent:*

(1) T *is a Fredholm operator with* index $(T) = 0$.
(2) T *is a compact perturbation of an invertible operator,*
i.e. T *can be written as the sum* $T = S + K$ *of an invertible operator* $S : X \to Y$ *and a compact operator* $K : X \to Y$.

Proof. (1) $\Rightarrow$ (2) Let T be a Fredholm operator with index $(T) = 0$.

By Theorem 6.3.6 we know that T can be given a matrix description

$$T = \begin{bmatrix} T_{11} & T_{12} \\ T_{21} & T_{22} \end{bmatrix} : \begin{bmatrix} X_1 \\ X_2 \end{bmatrix} \to \begin{bmatrix} Y_1 \\ Y_2 \end{bmatrix} ,$$

with respect to direct sum decompositions $X = X_1 \oplus X_2$ and $Y = Y_1 \oplus Y_2$ of the Banach spaces X and Y into closed linear subspaces, where X_2 and Y_2 are finite dimensional. Furthermore, $T_{11} : X_1 \to Y_1$ is a bijective operator with a bounded inverse operator, T_{12} and T_{22} are finite rank operators, and index $(T) = \dim(X_2) - \dim(Y_2)$.

Since index $(T) = 0$, the finite dimensional spaces X_2 and Y_2 have the same dimension, and hence we can choose an isomorphism $K_{22} : X_2 \to Y_2$.

With respect to the decompositions $X = X_1 \oplus X_2$ and $Y = Y_1 \oplus Y_2$, we now define the compact operator K by the matrix description

$$K = \begin{bmatrix} 0 & T_{12} \\ T_{21} & T_{22} - K_{22} \end{bmatrix} : \begin{bmatrix} X_1 \\ X_2 \end{bmatrix} \to \begin{bmatrix} Y_1 \\ Y_2 \end{bmatrix} .$$

And the invertible operator S by the matrix description

$$S = \begin{bmatrix} T_{11} & 0 \\ 0 & K_{22} \end{bmatrix} : \begin{bmatrix} X_1 \\ X_2 \end{bmatrix} \to \begin{bmatrix} Y_1 \\ Y_2 \end{bmatrix} .$$

Then we have $T = S + K$ as should be proved.

(2) $\Rightarrow$ (1) Next suppose that $T = S + K$ for an invertible operator S and a compact operator K. Since an invertible operator S is a Fredholm operator

with index $(S) = 0$ and since K is compact, we get by Proposition 6.3.8 that T is a Fredholm operator with index $(T) = 0$, as should be proved.   □

For any pair of Banach spaces X and Y, we denote by $\mathcal{F}_k(X, Y)$ the open subset of $\mathcal{F}(X, Y)$ containing the Fredholm operators of index $k \in \mathbb{Z}$. The subset $\mathcal{F}_k(X, Y)$ is not necessarily connected and may for some Banach spaces have infinitely many connectedness components. The 'trouble' is caused by the subset $\mathcal{F}_0(X, Y)$, which is closely related to the space of invertible operators $\mathcal{I}(X, Y)$, that may have infinitely many components; see Remark 6.4.5.

There are, however, important classes of Banach spaces where the connected components in the set of Fredholm operators are classified by the index. We finish this section with the primary example of this situation.

Theorem 6.4.13. *Let H be an infinite dimensional, separable Hilbert space. Then the set of Fredholm operators on H has the following properties:*

(1) The set of Fredholm operators $\mathcal{F}_0(H)$ of index zero is connected.
(2) The set of Fredholm operators $\mathcal{F}(H)$ can be given the structure of a monoid with composition of operators as the (associative) product.
(3) The index function $\mathcal{F}(H) \to \mathbb{Z}$ is a surjective homomorphism of $\mathcal{F}(H)$, with monoid structure, onto the integers $\mathbb{Z}$, with additive structure.
(4) The index function induces a bijective correspondence between the set of connected components of $\mathcal{F}(H)$ and $\mathbb{Z}$.

Proof.   **Assertion (1)** Follows by Proposition 6.4.12 and Theorem 6.4.6, since the linear space $\mathcal{C}(H)$ is connected.
Assertion (2) Follows immediately from Proposition 6.3.9.
Assertion (3) The homomorphism part follows from Proposition 6.3.9. That the index function is surjective follows from Example 6.3.10.
Assertion (4) Follows since the index function is surjective and since it has kernel $\mathcal{F}_0(H)$.   □

6.5   Integral operators and integral equations

Let $R^n = [a_1, b_1] \times \cdots \times [a_n, b_n]$ be a (generalized) closed interval in the real number space $\mathbb{R}^n$, given as the product of $n \geq 1$ closed intervals in $\mathbb{R}$. The Riemann integral of a continuous function in n real variables over a closed interval R^n can be introduced in complete analogy to the Riemann integral of a continuous function in one variable given in § 2.4.4.

Denote by $C(R^n)$ the Banach space of continuous $\mathbb{K}$-valued ($\mathbb{K} = \mathbb{R}, \mathbb{C}$)

functions $f : R^n \to \mathbb{K}$ equipped with the uniform norm

$$||f||_\infty = \sup_{x \in R^n} |f(x)| .$$

Let $\Phi = \Phi(x, y) : R^n \times R^m \to \mathbb{K}$ be a continuous function in two variables $(x, y) \in R^n \times R^m \subset \mathbb{R}^{n+m}$.

For $f \in C(R^n)$ we define the function $f_\Phi = f_\Phi(y) : R^m \to \mathbb{K}$ by

$$f_\Phi(y) = \int_{R^n} \Phi(x, y) f(x) dx .$$

Then we have

$$|f_\Phi(y_1) - f_\Phi(y_2)| \le \left(\int_{R^n} |\Phi(x, y_1) - \Phi(x, y_2)| \, dx \right) ||f||_\infty ,$$

for all $y_1, y_2 \in R^m$. Using that $\Phi = \Phi(x, y) : R^n \times R^m \to \mathbb{K}$ is uniformly continuous in the compact set $R^n \times R^m$, it is easy to prove that the function $f_\Phi = f_\Phi(y) : R^m \to \mathbb{K}$ is continuous, such that $f_\Phi \in C(R^m)$.

Hence we can define the mapping

$$T : C(R^n) \to C(R^m) \ \text{ by } \ T(f) = f_\Phi .$$

Clearly T is a linear operator, since integration is linear. For obvious reasons, the linear operator T is called an *integral operator* with *continuous kernel* Φ.

We shall now prove that T is a bounded linear operator.

First observe that the function $\phi = \phi(y) : R^m \to \mathbb{R}$ defined by

$$\phi(y) = \int_{R^n} |\Phi(x, y)| \, dx ,$$

is continuous in the compact set R^m. Therefore ϕ has a maximum value $\phi(y_0)$ attained in a point $y_0 \in R^m$, i.e.

$$\phi(y_0) = \int_{R^n} |\Phi(x, y_0)| \, dx = \sup_{y \in R^m} \int_{R^n} |\Phi(x, y| \, dx .$$

That T is a bounded linear operator now follows by the computations

$$|f_\Phi(y)| \le \int_{R^n} |\Phi(x, y) f(x)| \, dx \le \left(\int_{R^n} |\Phi(x, y)| \, dx \right) ||f||_\infty$$

$$\le \left(\sup_{y \in R^m} \int_{R^n} |\Phi(x, y)| \, dx \right) ||f||_\infty = \phi(y_0) ||f||_\infty ,$$

which shows that

$$||T(f)||_\infty = ||f_\Phi||_\infty \le \phi(y_0) ||f||_\infty .$$

We also get that the operator norm of T satisfies

$$\|T\| \leq \phi(y_0) = \sup_{y \in R^m} \int_{R^n} |\Phi(x,y)| \, dx \; .$$

With the intention to prove $\phi(y_0) \leq \|T\|$, we consider for every $\varepsilon > 0$ the function f_ε defined by

$$f_\varepsilon(x) = \frac{\overline{\Phi(x, y_0)}}{|\Phi(x, y_0)| + \varepsilon} \quad \text{for} \quad x \in R^n \; .$$

Conjugation is needed when $\mathbb{K} = \mathbb{C}$.

Then $\|f_\varepsilon\|_\infty \leq 1$ and we have the computation

$$\|T\| \geq \|T(f_\varepsilon)\|_\infty \geq |T(f_\varepsilon)(y_0)|$$

$$= \int_{R^n} \frac{|\Phi(x,y_0)|^2}{|\Phi(x,y_0)| + \varepsilon} \, dx \geq \int_{R^n} \frac{|\Phi(x,y_0)|^2 - \varepsilon^2}{|\Phi(x,y_0)| + \varepsilon} \, dx$$

$$= \int_{R^n} |\Phi(x,y_0)| \, dx - \varepsilon \mu(R^n) = \phi(y_0) - \varepsilon \mu(R^n) \; ,$$

where $\mu(R^n)$ is the Lebesgue measure of R^n, i.e. $\mu(R^n) = \prod_{i=1}^{n}(b_i - a_i)$.

Now $\phi(y_0) \leq \|T\|$, since $\|T\| \geq \phi(y_0) - \varepsilon \mu(R^n)$ for all $\varepsilon > 0$, and therefore

$$\|T\| = \phi(y_0) = \sup_{y \in R^m} \int_{R^n} |\Phi(x,y| \, dx \; .$$

The definition and main properties of integral operators with continuous kernel are listed in the following theorem.

Theorem 6.5.1. *Let $R^n \subset \mathbb{R}^n$ and $R^m \subset \mathbb{R}^m$ be (generalized) closed intervals in the real number spaces $\mathbb{R}^n$ and $\mathbb{R}^m$. Let $\Phi = \Phi(x,y) : R^n \times R^m \to \mathbb{K}$ be a continuous function in two variables $(x,y) \in R^n \times R^m \subset \mathbb{R}^{n+m}$. Then the operator $T : C(R^n) \to C(R^m)$ defined by*

$$T(f)(y) = \int_{R^n} \Phi(x,y) f(x) dx \quad for \quad f \in C(R^n), \; y \in R^m,$$

is a linear operator, called an integral operator *with continuous kernel Φ.*

Integral operators have the following main properties:

i) An integral operator T with continuous kernel is bounded with norm

$$\|T\| = \sup_{y \in R^m} \int_{R^n} |\Phi(x,y| \, dx \; .$$

ii) An integral operator T with continuous kernel is compact.

Proof. Everything has been proved except ii).

Let $\mathcal{A}$ be a family of maps forming a bounded subset in $C(R^n)$, i.e. there is a constant $c_1 > 0$ such that $\|f\|_\infty \le c_1$ for all $f \in \mathcal{A}$. We have to prove that the family of maps $T(\mathcal{A})$ is a compact subset in $C(R^m)$.

The continuous function $|\Phi| = |\Phi(x, y)|$ is bounded by a constant $c_2 > 0$ in the compact set $R^n \times R^m$. Hence we have

$$|T(f)(y)| \le \int_{R^n} |\Phi(x, y)| \, |f(x)| \, dx \le c_1 c_2 \mu(R^n) \,,$$

for all $f \in \mathcal{A}$ and all $y \in R^m$. This proves that $\|T(f)\|_\infty \le c_1 c_2 \mu(R^n)$ for all $f \in \mathcal{A}$, and hence that $T(\mathcal{A})$ is a bounded subset in $C(R^m)$.

From earlier we have the computation

$$|T(f)(y_1) - T(f)(y_2)| \le \left(\int_{R^n} |\Phi(x, y_1) - \Phi(x, y_2)| \, dx \right) \|f\|_\infty$$

for all $y_1, y_2 \in R^m$.

Using that $\Phi = \Phi(x, y) : R^n \times R^m \to \mathbb{K}$ is uniformly continuous in the compact set $R^n \times R^m$, we can for any given $\varepsilon > 0$ choose a $\delta > 0$, such that

$$|\Phi(x, y_1) - \Phi(x, y_2)| \le \frac{\varepsilon}{c_1 \mu(R^n)}$$

for all $x \in R^n$ and $\|y_1 - y_2\| \le \delta$.

Combining the above we get for all $f \in \mathcal{A}$,

$$|T(f)(y_1) - T(f)(y_2)| \le \varepsilon \quad \text{for} \quad \|y_1 - y_2\| \le \delta \,,$$

proving that the family of maps $T(\mathcal{A})$ is equicontinuous.

Since the family of maps $T(\mathcal{A})$ is bounded and equicontinuous, we get by the Ascoli-Arzela Theorem 1.3.14 that $T(\mathcal{A})$ is a compact subset in $C(R^m)$, which should be proved. $\square$

For each $p \ge 1$, we can define the space $L^p(R^n)$ of pth power (absolute) Lebesgue integrable functions on R^n as the completion of the space of continuous functions $C(R^n)$ with respect to the p-norm

$$\|f\|_p = \left(\int_{R^n} |f(x)|^p \, dx \right)^{1/p} \quad \text{for} \quad f \in C(R^n) \,.$$

The construction proceeds in complete analogy with the construction of the L^p-spaces in § 2.4, and all results are valid also in this more general setting.

Theorem 6.5.1 can be extended to all L^p-spaces on (generalized) closed intervals $R^n \subset \mathbb{R}^n$ and $R^m \subset \mathbb{R}^m$.

Theorem 6.5.2. *Let $R^n \subset \mathbb{R}^n$ and $R^m \subset \mathbb{R}^m$ be (generalized) closed intervals in the real number spaces $\mathbb{R}^n$ and $\mathbb{R}^m$. Let $\Phi = \Phi(x, y) : R^n \times R^m \to \mathbb{K}$ be a continuous function in two variables $(x, y) \in R^n \times R^m \subset \mathbb{R}^{n+m}$. Then the operator $T : L^p(R^n) \to L^p(R^m)$ defined by*

$$T(f)(y) = \int_{R^n} \Phi(x, y) f(x) dx \quad for \quad f \in L^p(R^n), \ y \in R^m,$$

is a compact, bounded linear operator with operator norm

$$\|T\| = \sup_{y \in R^m} \int_{R^n} |\Phi(x, y)| \, dx \ .$$

The version of an integral operator with continuous kernel in Theorem 6.5.2, follows easily from Theorem 6.5.1 by approximating the elements in $L^p(R^n)$ with continuous functions. The proof that the operator is compact exploits that the L^p-norm in $C(R^n)$ is dominated by the uniform norm.

The case $p = 2$ is of course of special interest. Proceeding in complete analogy with Example 3.1.6, we can define an inner product in $L^2(R^n)$, which turns this space into a separable Hilbert space, making all the machinery developed for separable Hilbert spaces in earlier chapters available.

The operator $T : L^2(R^n) \to L^2(R^m)$ defined in Theorem 6.5.2 is the classical example of a *Hilbert-Schmidt operator* with continuous kernel $\Phi = \Phi(x, y)$. Theorem 6.5.2 remains true also in case the kernel $\Phi \in L^2(R^n \times R^m)$.

We turn now to integral equations. In the most general terms, an integral equation is an equation in which an unknown function appears under an integral sign. There is a close connection between differential and integral equations, and some problems may be formulated either way as it is the case for example for the famous Maxwell's equations for electromagnetism.

Most of the classical integral equations studied in the physical sciences carry linearity in their formulations and are therefore closely related to linear integral operators.

The term integral equation was first used by the German mathematician Paul du Bois-Reymond (1831–1889) in a paper of 1888. However, it was the pioneering work of Volterra and not least the paper by Fredholm *"Sur une classe d'équations fonctionelles"*, published in final form 1903 in Acta Mathematica, that pushed the subject forward and stimulated the development of functional analysis. In his work, Fredholm introduced two types of integral equations that now carry his name.

We formulate the two types of Fredholm integral equations for the Banach space $C(R^n)$ in the notation from Theorem 6.5.1, assuming the closed intervals $R^n = R^m$ are identical, in particular that $n = m$. There are similar formulations for the Banach space $L^p(R^n)$ in the notation from Theorem 6.5.2.

(1) A *Fredholm integral equation of the first kind* has the form

$$\int_{R^n} \Phi(x,y)f(x)dx = g(y),$$

for $f, g \in C(R^n)$ and $x, y \in R^n$. As operator equation

$$T(f) = g \quad \text{for} \quad f, g \in C(R^n).$$

(2) A *Fredholm integral equation of the second kind* has the form

$$\int_{R^n} \Phi(x,y)f(x)dx - \lambda f(y) = g(y),$$

for $f, g \in C(R^n)$, $x, y \in R^n$ and $\lambda \in \mathbb{K}$, $\lambda \neq 0$. As operator equation

$$(T - \lambda I)(f) = g \quad \text{for} \quad f, g \in C(R^n), \ \lambda \neq 0,$$

where I denotes the identity operator in $C(R^n)$.

6.6 Operator equations of Fredholm Type

For any type of equation in mathematics, the main problems concern existence, uniqueness, and number of solutions to the equation, referred to as questions about *solvability* of the equation.

In the study of solvability of the Fredholm integral equations of first and second kind, introduced in the previous section, it turns out that we can substitute the integral operator in the equations with a compact linear operator $T : X \to X$ on an arbitrary Banach space X over $\mathbb{K} = \mathbb{R}, \mathbb{C}$. Since this requires no more work, we shall do this in the following.

If I denotes the identity operator in the Banach space X, and $\lambda \in \mathbb{K}$ is a non-zero number, we shall in other words examine the solvability of the two types of operator equations defined as follows.

(1) A *Fredholm equation of the first kind in a Banach space* is given by

$$T(f) = g \quad \text{for} \quad f, g \in X,$$

where T is a compact linear operator on a Banach space X.

(2) *A Fredholm equation of the second kind in a Banach space* is given by

$$(T - \lambda I)(f) = g \quad \text{for} \quad f, g \in X, \ \lambda \neq 0,$$

where T is a compact linear operator on a Banach space X.

The optimal situation in relation to solvability of an equation with fixed left-hand side is that it has a unique solution for every right-hand side of the equation. For Fredholm equations of the first kind this situation is rarely the case, and it is only a possibility for finite dimensional Banach spaces.

Theorem 6.6.1. *For the Fredholm equation $T(f) = g$ of the first kind, defined by a compact linear operator T on a Banach space X, it holds that if the equation has a unique solution $f \in X$ for every $g \in X$, then the Banach space X must be finite dimensional.*

Proof.   If the equation $T(f) = g$ has a unique solution $f \in X$ for every $g \in X$, then $T : X \to X$ is a bijective, bounded linear operator between Banach spaces, and hence T is invertible by Banach's Theorem 1.8.3. Since T is compact, it maps the closed unit ball C in X into a compact subset $T(C)$ of X, which is next mapped into a compact set by the continuous inverse operator T^{-1}. Hence the closed unit ball $C = T^{-1}(T(C))$ in X is compact, and therefore X has finite dimension by Theorem 4.2.3.   □

For Fredholm equations of the second kind there are better possibilities for existence and uniqueness of solutions, since the operator in the equation in many cases is invertible. It all depends on the parameter $\lambda \neq 0$ in the equation.

We finish this chapter with the following main theorem on Fredholm equations of the second kind, which has become known as the Fredholm alternative.

Theorem 6.6.2 (The Fredholm Alternative). *Let T be a compact linear operator on the Banach space X, and let $\lambda \in \mathbb{K}$ be a non-zero number.*

For the Fredholm equation of the second kind

$$(T - \lambda I)(f) = g \quad for \quad f, g \in X, \ \lambda \neq 0,$$

one and only one of the following two statements is true.

(1) *The inhomogeneous equation*

$$(T - \lambda I)(f) = g$$

has a unique solution $f \in X$ for each $g \in X$.

(2) The homogeneous equation

$$(T - \lambda I)(f) = 0$$

has non-zero solutions (hence λ is an eigenvalue of T), and the set of solutions to the homogeneous equation span a finite dimensional subspace of X (the eigenspace for λ).

Proof. Since T is a compact operator, it follows from Proposition 6.3.8 that $T - \lambda I$ is a Fredholm operator with $\text{index}(T - \lambda I) = \text{index}(\lambda I) = 0$. This gives the equality $\dim(\ker(T - \lambda I)) = \dim(\text{coker}(T - \lambda I))$.

From this follows that the operator $T - \lambda I$ is surjective (equivalently $\text{coker}(T - \lambda I) = 0$) if and only if it is injective (equivalently $\ker(T - \lambda I) = 0$). This implies that $T - \lambda I$ is surjective if and only if it is bijective, and hence by Banach's Theorem 1.8.3, if and only if it is invertible.

The theorem now follows by observing that statement (1) is equivalent to the operator $T - \lambda I$ being surjective, and statement (2) to the operator $T - \lambda I$ being not surjective/injective. For (2) we use also that the dimension of $\ker(T - \lambda I)$ is finite. $\square$

Exercises

Basic prerequisites for the exercises from topology, including completeness of metric spaces (Banach fixed point theorem), and the theory of normed vector spaces, including the operator norm of a bounded linear mapping, can all be found in the general reference V.L. Hansen: *Fundamental Concepts in Modern Analysis*, World Scientific, 1999. A short summary of the necessary results from metric topology is given in Chapter 1.

Chapter 1

Exercise 1. Consider the metric space (M, d) where $M = [1, \infty)$ and d is the usual distance in $\mathbb{R}$. Define the mapping $T : M \to M$ by

$$Tx = \frac{x}{2} + \frac{1}{x} \quad \text{for } x \in M .$$

Show that T is a contraction and determine the minimal contraction factor λ. Determine the fixed point for T.

Exercise 2. A mapping T of a metric space (M, d) into itself is called a *weak contraction* if

$$d(Tx, Ty) < d(x, y) \quad \text{for all } x, y \in M, x \neq y .$$

1) Show that T has at most one fixed point.
2) Show that T does not necessarily have a fixed point.

<u>Hint:</u> Consider the function $Tx = x + \frac{1}{x}$ for $x \geq 1$.

Exercise 3. Let T be a mapping from a complete metric space (M, d) into itself, and assume that there is a natural number m such that T^m is a contrac-

tion. Show that T has one and only one fixed point.

Exercise 4. In mathematics one often considers iteration schemes of the form

$$x_n = g(x_{n-1})$$

for a real-valued, differentiable function g of class C^1.

Show that the sequence (x_n) is convergent for any choice of x_0 if there is a real number α in the interval $0 < \alpha < 1$, such that

$$|g'(x)| \leq \alpha \quad \text{for all } x \in \mathbb{R} .$$

Exercise 5. A general method to approximate the solution to an equation is to try to bring the equation into the form $x = g(x)$, and then choose an x_0, and use the iteration scheme $x_n = g(x_{n-1})$. Assume that g is a C^1-function in the interval $[x_0 - \delta, x_0 + \delta]$, and that $|g'(x)| \leq \alpha < 1$ for $x \in [x_0 - \delta, x_0 + \delta]$, and moreover

$$|g(x_0) - x_0| \leq (1 - \alpha)\delta .$$

Show that there is one and only one solution $x \in [x_0 - \delta, x_0 + \delta]$ to the equation, and that $x_n \to x$.

Exercise 6. Solve by iteration the equation $f(x) = 0$ for $f \in C^1([a,b])$, where $f(a) < 0 < f(b)$ and f' is bounded and strictly positive in $[a, b]$.
Hint: Take $g(x) = x - \lambda f(x)$ for a suitable choice of λ.

Exercise 7. Show that the equation $f(x) = x^3 + x - 1 = 0$ can be solved by the iteration

$$x_n = g(x_{n-1}) = 1/(1 + x_{n-1}^2) .$$

Find x_1, x_2, x_3 for $x_0 = 1$, and find an estimate for $|x - x_n|$.

Exercise 8. Show that any finite dimensional linear subspace U in a normed vector space V is a closed set in V.

Exercise 9. Let $f : X \to Y$ be an arbitrary mapping between topological spaces X and Y. In the product space $X \times Y$ equipped with the product topology, consider the *graph* $G(f)$ of f, i.e. the subset

$$G(f) = \{(x, f(x)) \,|\, x \in X\} .$$

1) Suppose that Y is a Hausdorff space. Prove that the graph $G(f)$ of f is a closed subset in $X \times Y$ if $f : X \to Y$ is a continuous mapping.

(A topological space Y is called a *Hausdorff space* if, for every pair of points $y_1, y_2 \in Y$ with $y_1 \neq y_2$, there exists a corresponding pair of mutually disjoint, open sets U_1 and U_2 in Y, such that $y_1 \in U_1$ and $y_2 \in U_2$.)

2) Is it necessary that Y is a Hausdorff space for 1) to hold?

<u>Hint:</u> Consider the graph of the identity mapping $1_X : X \to X$.

Exercise 10. Show that a closed linear subspace of a Banach space is itself a Banach space.

Exercise 11. Let $(V, ||\cdot||_V)$ and $(W, ||\cdot||_W)$ be normed vector spaces. Define the *product* $V \otimes W$ of normed vector spaces by giving the set

$$V \otimes W = \{(x, y) \,|\, x \in V, y \in W\} \,,$$

the obvious entrywise defined vector space structure and the norm

$$||(x, y)||_1 = ||x||_V + ||y||_W \quad \text{for} \quad x \in V, \ y \in W \,.$$

Now assume that $(V, ||\cdot||_V)$ and $(W, ||\cdot||_W)$ are Banach spaces.

1) Prove that $(V \otimes W, ||\cdot||_1)$ is a Banach space.

2) Prove that the graph $G(f)$ of a continuous, linear mapping $f : V \to W$ is a closed linear subspace of $(V \otimes W, ||\cdot||_1)$.

3) Prove that the graph $G(f)$ of a continuous, linear mapping $f : V \to W$ is a Banach space.

Exercise 12. Let $(V, ||\cdot||_V)$ and $(W, ||\cdot||_W)$ be normed vector spaces, and let $\mathcal{E}$ be an equicontinuous family of linear operators $T : V \to W$.

Prove that for any bounded subset $E \subset V$, there exists a bounded subset $F \subset W$ such that $T(E) \subset F$ for every $T \in \mathcal{E}$.

Exercise 13. (The Banach-Steinhaus Theorem) Let $(V, ||\cdot||_V)$ and $(W, ||\cdot||_W)$ be normed vector spaces with V a Banach space. Suppose that $\mathcal{E}$ is a family of bounded linear operators $T : V \to W$ for which the set

$$\mathcal{E}(x) = \{T(x) \,|\, T \in \mathcal{E}\}$$

is a bounded subset in W for all $x \in V$.

Prove that $\mathcal{E}$ is an equicontinuous family of maps.

Exercise 14. Let $(V, ||\cdot||)$ be an infinite dimensional normed vector space, which is the union of countably many finite dimensional subspaces.

Prove that V cannot be a Banach space in the norm $||\cdot||$.

Definition. Let $T : D(T) \subset V \to W$ be a linear operator between normed vector spaces $(V, ||\cdot||_V)$ and $(W, ||\cdot||_W)$ defined in the linear subspace $D(T)$ of V. Then T is called a *closed* linear operator if the graph $G(T)$ of T is a closed linear subspace of the product space $(V \oplus W, ||\cdot||_1)$; cf. Exercise 11.

Exercise 15. Prove that a linear operator $T : D(T) \subset V \to W$ between normed vector spaces $(V, ||\cdot||_V)$ and $(W, ||\cdot||_W)$ is a closed linear operator if and only if for each sequence (x_n) in the domain of definition $D(T)$ of T that converges to a point $x \in V$, and for which the image sequence (Tx_n) converges to a point $y \in W$, the limit point x actually belongs to $D(T)$ and $Tx = y$.

Exercise 16. Prove that a linear operator $T : V \to W$ between Banach spaces $(V, ||\cdot||_V)$ and $(W, ||\cdot||_W)$ is bounded if and only if it is closed.

Exercise 17. Let $T : D(T) \subset V \to W$ be a closed linear operator between two normed vector spaces defined in the linear subspace $D(T)$ of V. Show that $\ker(T)$ is a closed subspace of V.

Exercise 18. Let V be a Banach space and let T be a closed linear operator $T : D(T) \subset V \to V$ defined in the linear subspace $D(T)$ of V.

Prove that $A+T$ and TA are closed linear operators for any bounded linear operator $A : V \to V$.

Chapter 2

Exercise 19. Let φ be a real-valued continuous function defined for $x \geq 0$, and assume that $\lim_{x \to \infty} \varphi(x)$ exists (and is finite). Show that for every $\epsilon > 0$ there exists an integer $n \in \mathbb{N}$ and real constants $a_k, k = 0, 1, ..., n$, such that

$$|\varphi(x) - \sum_{k=0}^{n} a_k e^{-kx}| \leq \epsilon \quad \text{for all} \quad x \geq 0 .$$

Exercise 20. Prove the inequalities of Hölder, Cauchy-Schwarz and Minkowski for the spaces of absolute p^{th}-power Riemann integrable functions with compact support $\mathcal{R}_0^p(\mathbb{R})$.

Exercise 21. Prove that if (f_n) is a Cauchy sequence in $(\mathcal{R}_0^p(\mathbb{R}), ||\cdot||_p)$, then there exists a Cauchy sequence (g_n) in $(C_0(\mathbb{R}), ||\cdot||_p)$ such that $||f_n - g_n||_p \to 0$ for $n \to \infty$.

Exercise 22. For an arbitrary real number $h \in \mathbb{R}$, define the operator τ_h on $L^2(\mathbb{R})$ by

$$\tau_h f(x) = f(x - h) \quad \text{for } f \in L^2(\mathbb{R}) .$$

Show that τ_h is bounded.

Exercise 23. Let c denote the set of convergent complex sequences (x_n). Show that c is a Banach space when equipped with the norm

$$||(x_n)||_\infty = \sup_{n \in \mathbb{N}} |x_n| .$$

<u>Hint:</u> Show that $(c, ||\cdot||_\infty)$ is a closed linear subspace in the Banach space of bounded complex sequences $(l^\infty, ||\cdot||_\infty)$.

Exercise 24. In the Banach space of bounded complex sequences $(l^\infty, ||\cdot||_\infty)$, we consider the subset c_0 consisting of the sequences converging to 0 and the subset c_{00} consisting of the sequences with only a finite number of elements different from 0.
　1) Show that c_0 and c_{00} are linear subspaces of l^∞.
　2) Investigate if c_0 and/or c_{00} are Banach spaces.

Exercise 25. Consider the *Volterra integral equation*:

$$x(t) - \mu \int_a^t k(t,s)x(s)ds = v(t), \quad t \in [a,b] ,$$

where $v \in C([a,b])$, $k \in C([a,b]^2)$ and $\mu \in \mathbb{C}$.
　Show that the equation has a unique solution $x \in C([a,b])$ for any $\mu \in \mathbb{C}$.

<u>Hint:</u> Write the Volterra integral equation in the form $x = Tx$ where

$$Tx = v(t) + \mu \int_a^t k(t,s)x(s)ds .$$

Take $x_0 \in C([a,b])$ and define by iteration $x_{n+1} = Tx_n$. Show by induction that

$$|T^m x(t) - T^m y(t)| \le |\mu|^m c^m \frac{(t-a)^m}{m!} d_\infty(x,y) ,$$

where $c = \max|k|$ and $d_\infty(x,y) = \sup_{t\in[a,b]}|x(t) - y(t)|$. Then show (by looking at $d_\infty(T^m x, T^m y)$) that T^m is a contraction for some m and argue that T must then have a unique fixed point in the metric space $(C([a,b]), d_\infty)$.

Exercise 26. Let $f \in L^1(\mathbb{R})$.
1) Can we conclude that $f(x) \to 0$ for $|x| \to \infty$?
2) Can we find $a, b \in \mathbb{R}$ such that $|f(x)| \le b$ for $|x| \ge a$?

Exercise 27. In the vector space $C([a,b])$ of real-valued, continuous functions in the interval $[a,b]$, consider the functions $e_0(t), e_1(t), ..., e_n(t)$, where $e_j(t)$ is a polynomial of degree j, for each $j = 0, 1, ..., n$.
Show that $e_0, e_1, ..., e_n$ are linearly independent in $C([a,b])$.

Exercise 28. Let U_1 and U_2 be subspaces of the vector space V. Show that $U_1 \cap U_2$ is a subspace in V. Is $U_1 \cup U_2$ always a subspace? If not, state conditions such that $U_1 \cup U_2$ is a subspace in V.

Exercise 29. Let V denote the set of all real $n \times n$-matrices.
Show that V is a vector space, when equipped with the obvious entrywise defined operations.
Is the set of all *regular* $n \times n$-matrices a subspace of V?
Is the set of all *symmetric* $n \times n$-matrices a subspace of V?

Exercise 30. In the space $C([a,b])$ of real-valued, continuous functions in the interval $[a,b]$, consider the sets
$U_1 =$ the set of polynomials defined on $[a,b]$.
$U_2 =$ the set of polynomials defined on $[a,b]$ of degree $\le n$.
$U_3 =$ the set of polynomials defined on $[a,b]$ of degree $= n$.
$U_4 =$ the set of all $f \in C([a,b])$ with $f(a) = f(b) = 0$.
$U_5 = C^1([a,b])$.
Which ones of the sets U_i, $i = 1, 2, ..., 5$ are subspaces of $C([a,b])$?

Exercise 31. In $C([-1,1])$ we consider the sets U_1 and U_2 consisting of the odd and the even functions in $C([-1,1])$, respectively.
Show that U_1 and U_2 are subspaces and that $U_1 \cap U_2 = \{0\}$.
Show that every $f \in C([-1,1])$ can be written in the form $f = f_1 + f_2$, where $f_1 \in U_1$ and $f_2 \in U_2$, and that this decomposition is unique.

Exercise 32. Consider the space $C^1([a,b])$ of real-valued, differentiable functions of class C^1 in the interval $[a, b]$ with the uniform norm

$$\|f\|_\infty = \sup_{t \in [a,b]} |f(t)| .$$

1) Prove that

$$\|f\|_\infty = \sup_{t \in (a,b)} |f(t)| .$$

2) Show that $(C^1([a, b]), \|\cdot\|_\infty)$ is not a Banach space.
3) Show that

$$\|f\|_\infty^* = \sup_{t \in [a,b]} |f(t)| + \sup_{t \in [a,b]} |f'(t)|$$

is also a norm in $C^1([a, b])$ and that $C^1([a, b])$ is a Banach space with this norm.

Exercise 33. Let $f \in C([a, b])$ and consider the p-norms

$$\|f\|_p = \left(\int_a^b |f(t)|^p dt \right)^{\frac{1}{p}}, \quad p \geq 1 ,$$

and the uniform norm

$$\|f\|_\infty = \sup_{t \in [a,b]} |f(t)| .$$

Show that $\|f\|_p \to \|f\|_\infty$ for $p \to \infty$.

Exercise 34. Let V be a normed vector space and let $x_1, \ldots, x_k$ be k linearly independent vectors from V. Show that there exists a positive constant m such that for all scalars $\alpha_i \in \mathbb{C}, \ i = 1, \ldots, k$, we have

$$\|\alpha_1 x_1 + \cdots + \alpha_k x_k\| \geq m(|\alpha_1| + \cdots + |\alpha_k|).$$

Exercise 35. Let T be a linear operator from a finite dimensional vector space into itself. Prove that T is injective if and only if T is surjective.

Exercise 36. Let $T : C^\infty(\mathbb{R}) \to C^\infty(\mathbb{R})$ be the linear mapping defined by $Tf = f'$.
1) Show that T is surjective.
2) Is T injective?

Exercise 37. Consider the Lebesgue spaces $L^1([0,1])$ and $L^2([0,1])$ on the unit interval $[0,1]$.

1) Define an inclusion mapping $I : L^2([0,1]) \to L^1([0,1])$ by $I(f) = fe$, where $e \in L^2([0,1])$ is the constant function with value 1. Show that I is a continuous linear mapping.

2) For each $n \in \mathbb{N}$ consider the subset

$$M_n = \{f \in L^2([0,1]) \,|\, \|f\|_1^2 \le n\} \quad \text{in} \quad L^2([0,1]) \,.$$

Prove that $I(M_n)$ is a closed subset in $L^1([0,1])$ with empty interior.

3) For each $n \in \mathbb{N}$ define the function

$$g_n(x) = \begin{cases} n \text{ for } 0 \le x \le n^{-3} \\ 0 \text{ for } n^{-3} < x \le 1 \,. \end{cases}$$

Show that for every $f \in L^2([0,1])$, but not for every $f \in L^1([0,1])$, it holds

$$\lim_{n\to\infty} \|f g_n\|_1 = 0 \,.$$

4) Show that the inclusion mapping $I : L^2([0,1]) \to L^1([0,1])$ is continuous, but not surjective.

Exercise 38. In l^∞, the vector space of bounded sequences, we consider the sets U_1 and U_2, where U_1 denotes the set of sequences with only a finite number of elements different from 0, and U_2 the set of sequences with all but the N first elements different from 0.

Are U_1 and/or U_2 closed subspaces in l^∞?

Are U_1 and/or U_2 finite dimensional?

Exercise 39. Let $T : V \to W$ be a linear operator between normed vector spaces V and W.

Show that the image $T(V)$ is a subspace of W.

Show that the kernel (or nullspace) $\ker(T)$ is a subspace of V.

If T is bounded, is it true that $T(V)$ and/or $\ker(T)$ are closed?

Exercise 40. In the Banach space l^p, for $1 \le p \le \infty$, we are given a sequence (x_n) converging to an element x, where

$$x_n = (x_{n1}, x_{n2}, \dots) \quad \text{and} \quad x = (x_1, x_2, \dots) \,.$$

Show that if $x_n \to x$ in l^p, then $x_{nk} \to x_k$ for all $k \in \mathbb{N}$.

If $x_{nk} \to x_k$ for all $k \in \mathbb{N}$, is it true that $x_n \to x$ in l^p?

Exercise 41. Prove that the vector space of finitely non-zero sequences is isomorphic to the vector space of all polynomial functions defined in a closed and bounded interval $[a, b]$ with $a < b$.

Exercise 42. Let T be a linear mapping from $\mathbb{R}^m$ to $\mathbb{R}^n$, both equipped with the 2-norm. Let (a_{ij}) denote a real $n \times m$ matrix corresponding to T. Show that T is a bounded linear operator with $||T||^2 \leq \sum_{i=1}^{m} \sum_{j=1}^{n} a_{ij}^2$.

Exercise 43. Let $[a, b]$ be an arbitrary closed and bounded interval. Show that $L^2([a, b]) \subset L^1([a, b])$.

Chapter 3

Exercise 44. Let $T : V \to W$ be an injective linear mapping of the vector space V into the vector space W. Let $(\cdot, \cdot)_W$ be an inner product in W.
   1) Show that we define an inner product $(\cdot, \cdot)_V$ in V by the definition

$$(x, y)_V = (Tx, Ty)_W \quad \text{for } x, y \in V .$$

   2) Show that T is a bounded linear operator, when V is equipped with the inner product $(\cdot, \cdot)_V$ and W with the inner product $(\cdot, \cdot)_W$. Determine the operator norm $||T||$ of T.

Exercise 45. Let H be a Hilbert space of infinite dimension. Show that H is separable by a basis if and only if it contains a countable dense subset.
   Does part of this result hold in greater generality than for Hilbert spaces?

Exercise 46. This exercise examines separability of the sequence spaces.
   1) Prove that the sequence spaces l^p, $p \geq 1$, are separable.
   2) Prove that the space l^∞ of bounded complex sequences is not separable.
Hint: First prove that the subset $\mathcal{B}$ of sequences (x_n) with all elements x_n either 0 or 1 is uncountable. Next prove (indirectly) that l^∞ does not contain a countable dense subset by exploiting that any two mutually different sequences in $\mathcal{B}$ have norm distance 1 and hence cannot both be contained in an open ball of radius $1/3$ in l^∞.

Exercise 47. Let $I = [a, b]$ be a bounded interval and consider the linear mapping T from $C([a,b])$ into itself, given by

$$Tf(t) = \int_a^t f(s)ds.$$

We assume that $C([a,b])$ is equipped with the uniform norm.
Show that T is bounded and find $\|T\|$.
Show that T is injective and find $T^{-1} : T(C([a,b])) \to C([a,b])$.
Is T^{-1} bounded?

Exercise 48. Let T be a bounded linear operator from a normed vector space V into a normed vector space W, and assume that T is surjective. Assume that there is a $c > 0$ such that

$$\|Tx\| \ge c\|x\| \quad \text{for all } x \in V .$$

Show that T^{-1} exists and that T^{-1} is bounded.

Exercise 49. Prove that in a *real* vector space with inner product we have

$$(x,y) = \frac{1}{4}(\|x+y\|^2 - \|x-y\|^2),$$

and in a *complex* vector space with inner product we have

$$(x,y) = \frac{1}{4}(\|x+y\|^2 - \|x-y\|^2 + i\|x+iy\|^2 - i\|x-iy\|^2).$$

These are the so-called *Polarization identities*. They show that, in a Hilbert space, the inner product is determined by the norm.

Exercise 50. Let V be a *real* normed vector space, and assume that the norm satisfies

$$\|x+y\|^2 + \|x-y\|^2 = 2(\|x\|^2 + \|y\|^2)$$

for all $x, y \in V$.
Show that

$$(x,y) = \frac{1}{4}(\|x+y\|^2 - \|x-y\|^2)$$

defines an inner product in V and that the norm is induced by this inner product.

Exercise 51. Prove that the uniform norm on the space of continuous functions $C([a,b])$ is not induced by an inner product.

Exercise 52. Prove that in a *real* vector space with inner product we have that $(x + y, x - y) = 0$ if $||x|| = ||y||$.

In the case $V = \mathbb{R}^2$, this corresponds to a well-known geometric statement. Which statement?

Exercise 53. Let V_i, $i = 1, \ldots, k$, be vector spaces equipped with inner products $(\cdot, \cdot)_i$, respectively. We define the product space $\bigotimes_{i=1}^{k} V_i$, in analogy with Exercise 11.

Show that we can define an inner product in $\bigotimes_{i=1}^{n} V_i$ by

$$((x_1, x_2, \ldots, x_k), (y_1, y_2, \ldots, y_k)) = \sum_{i=1}^{k} (x_i, y_i)_i,$$

and that $\bigotimes_{i=1}^{n} V_i$ with this inner product is a Hilbert space if all the inner product spaces V_i, $i = 1, \ldots, k$, are Hilbert spaces.

Exercise 54. Let x and y be vectors in a vector space with an inner product. Show that $(x, y) = 0$ if and only if

$$||x + \alpha y|| = ||x - \alpha y||$$

for all scalars α.

Moreover, show that $(x, y) = 0$ if and only if

$$||x + \alpha y|| \geq ||x||$$

for all scalars α.

Exercise 55. Let x and y be vectors in a *complex* vector space with an inner product, and assume that

$$||x + y||^2 = ||x||^2 + ||y||^2.$$

Does this imply that $(x, y) = 0$?

Exercise 56. Let V be a complex vector space with an inner product and assume that $T \in B(V)$.

Show that $(Tx, y) = 0$ for all $x, y \in V$ if and only if T is the zero operator.

Show next that $(Tx, x) = 0$ for all $x \in V$ if and only if T is the zero operator.

Are these results true if the vector space is a real vector space?

Exercise 57. Let T be a linear operator $T : L^2(\mathbb{R}) \to L^2(\mathbb{R})$ satisfying that $f \geq 0$ implies that $Tf \geq 0$.

Show that

$$||T(|f|)|| \geq ||Tf||$$

for all $f \in L^2(\mathbb{R})$.

Show that T is bounded.

Exercise 58. Consider the Hilbert space $H = L^2([-\pi, \pi])$. Using classical Fourier theory we can define an orthonormal basis (e_n) in H by

$$e_n(t) = \frac{1}{\sqrt{2\pi}} e^{int}, \quad n \in \mathbb{Z}.$$

[This follows, since it is known that (e_n) is an orthonormal basis in the space of smooth functions in $[-\pi, \pi]$, which is dense in H.]

For $f \in H$ we define the *Fourier transform* $\hat{f}$ by

$$\hat{f}(x) = \frac{1}{2\pi} \int_{-\pi}^{\pi} f(t) e^{-ixt} dt .$$

1) Show that $\hat{f}$ exists for all $x \in \mathbb{R}$.

2) Use the function $\hat{f}$ to express the Fourier expansion of $f \in H$ in terms of the *orthogonal* basis $(\sqrt{2\pi}e_n)$.

3) Let $\gamma \in \mathbb{R}$ and $f \in H$ be given and define the function g by

$$g(t) = f(t) e^{-i\gamma t} .$$

Find $\hat{g}(x)$.

4) Show that for any $\gamma \in \mathbb{R}$ and $f \in H$ we have

$$\sum_{n=-\infty}^{\infty} |\hat{f}(n + \gamma)|^2 = \frac{1}{2\pi} ||f||_2^2 .$$

5) Take $f = 1$ and $\gamma = \frac{\theta}{\pi}$, $\theta \notin \{p\pi \mid p \in \mathbb{Z}\}$ and show that

$$\frac{1}{\sin^2(\theta)} = \sum_{n=-\infty}^{\infty} \frac{1}{(n\pi + \theta)^2} .$$

Exercise 59. Let $H = L^2([0, 1])$, and consider the operator

$$Tf(x) = \sqrt{3} x f(x^3) .$$

1) Show that $T \in B(H)$ and find $||T||$.

2) Show that T^{-1} exists and that $T^{-1} \in B(H)$. Determine $T^{-1}g(y)$ for $g \in H$, and find $||T^{-1}||$.

Exercise 60. Let M be a subset of a Hilbert space H. Show that $M^{\perp}$ is a closed subspace of H.

Show that $M \subseteq (M^{\perp})^{\perp}$, and show that $(M^{\perp})^{\perp}$ is the smallest closed subspace containing M.

Exercise 61. Let (x_n) be an orthogonal sequence in a Hilbert space H, satisfying that

$$\sum_{n=1}^{\infty} ||x_n||^2 < \infty .$$

Show that the series $\sum_{n=1}^{\infty} x_n$ is convergent in H.

Is this still true if we drop the orthogonality assumption?

Exercise 62. Let H be an infinite dimensional, separable Hilbert space. Show that there is a sequence of vectors (x_n) such that $||x_n|| = 1$ for all n, and such that $(x_n, x) \to 0$ for all $x \in H$.

Exercise 63. Let H be a Hilbert space. Show that

$$||x - z|| = ||x - y|| + ||y - z||$$

if and only if $y = \alpha x + (1 - \alpha)z$ for some $\alpha \in [0, 1]$.

Exercise 64. Let (e_n) be an orthonormal basis for $L^2([0,1])$. Construct an orthonormal basis for $L^2([a, b])$, where $[a, b]$ is an arbitrary closed and bounded interval.

Exercise 65. Let (e_n) be an orthonormal sequence in the space of square integrable functions $L^2([a, b])$ in a closed and bounded interval $[a, b]$, with the property that for any *continuous* function $f \in L^2([a, b])$ and any $\epsilon > 0$ we can find an integer $N \in \mathbb{N}$ and constants $a_1, a_2, \ldots, a_N$ such that

$$||f - \sum_{k=1}^{N} a_k e_k||_2 < \epsilon .$$

Show that (e_n) is an orthonormal basis for $L^2([a, b])$.

Exercise 66. Construct the sequence of *Haar-functions* (h_n) in $L^2([0,1])$ by defining $h_1(t) = 1$, for $t \in [0,1]$, and h_n for $n \geq 2$ by

$$
h_{2^m+k}(t) = \begin{cases}
\sqrt{2^m} & \text{for} \quad \frac{k-1}{2^m} \leq t \leq \frac{2k-1}{2^{m+1}} \\
-\sqrt{2^m} & \text{for} \quad \frac{2k-1}{2^{m+1}} \leq t \leq \frac{k}{2^m} \\
0 & \text{else,}
\end{cases}
$$

for $k = 1, 2, \ldots, 2^m$ and $m = 0, 1, 2, \ldots$.
    1) Sketch the graphs of $h_1, h_2, \ldots, h_8$.
    2) Show that (h_n) is an orthonormal sequence in $L^2([0,1])$.
    3) Show that (h_n) is an orthonormal basis in $L^2([0,1])$.

Exercise 67. Let H be a Hilbert space and let P and Q denote the orthogonal projections onto the closed subspaces M and N, respectively. Show that if $M \perp N$, then $P + Q$ is the orthogonal projection onto $M \oplus N$.

Exercise 68. Let P and Q denote orthogonal projections in a Hilbert space, and assume that $PQ = QP$. Show that $P + Q - PQ$ is an orthogonal projection and find the image of $P + Q - PQ$.

Exercise 69. Let φ denote a linear functional on a vector space V, and assume that $\ker(\varphi) \neq V$. Let $x_0 \in V \setminus \ker(\varphi)$. Show that any vector $x \in V$ can be written in the form $x = ax_0 + y$, where $y \in \ker(\varphi)$. Is this expansion unique?

Exercise 70. Let φ and ψ denote linear functionals on a vector space V, and assume that $\ker(\varphi) = \ker(\psi)$.
    Show that there is a constant $\alpha \in \mathbb{C}$ such that $\alpha\varphi = \psi$.

Exercise 71. Let V and W be Hilbert spaces and let $T : V \to W$ be a bounded linear operator. Show that the image of a weakly convergent sequence in V is a weakly convergent sequence in W.

Exercise 72. Let V be a normed vector space. Show that no pair of operators $S, T \in B(V)$ satisfies the *canonical commutator relation*:

$$
[S, T] = ST - TS = I \ .
$$

<u>Hint:</u> Show by induction that $ST^n - T^n S = nT^{n-1}$, $n \in \mathbb{N}$, and use this to estimate $\|S\|$ and $\|T\|$.

Exercise 73. Let (e_n) denote an orthonormal basis in a Hilbert space H, and define the operator T by

$$T(\sum_{k=1}^{\infty} x_k e_k) = \sum_{k=1}^{\infty} x_k e_{k+1}.$$

Show that $T \in B(H)$ and find $\|T\|$.
Show that T is injective and find T^{-1}.

Exercise 74. In the Hilbert space of square summable sequences l^2, define the operator

$$T((x_n)) = (y_n) \ ,$$

by setting

$$y_1 = x_1 \quad \text{and} \quad y_n = \frac{1}{2^{n-1}}(x_1 + x_2 + \cdots + x_n) \quad \text{for } n \geq 2 \ .$$

Show that the operator T is bounded and not surjective.

Exercise 75. Let (e_k) be an orthonormal basis in a Hilbert space H, and let $T \in B(H)$. Define for $j, k \in \mathbb{N}$ the numbers

$$t_{jk} = (Te_j, e_k) \ .$$

Show that

$$Te_j = \sum_{k=1}^{\infty} t_{jk} e_k \ ,$$

and that $\sum_{k=1}^{\infty} |t_{jk}|^2 < \infty$ for $j \in \mathbb{N}$.
    The matrix (t_{jk}) is called the *matrix form* for T with respect to the orthonormal basis (e_k).
    Let $A, B \in B(H)$ have the forms (a_{jk}) and (b_{jk}), respectively. Find the forms for $A + B$ and AB.

Chapter 4

Exercise 76. Let V denote a normed vector space with norm $\|\cdot\|$. The vector space of bounded linear functionals $\varphi : V \to \mathbb{C}$ on V is called the *dual space* of V and is denoted by V^*. Introduce the *dual norm* $\|\cdot\|^*$ in V^* by taking $\|\varphi\|^*$ to be the operator norm of the bounded linear functional $\varphi : V \to \mathbb{C}$.

Let $x \in V$. Show that

$$g_x(\varphi) = \varphi(x), \quad \varphi \in V^* \ ,$$

defines an element $g_x \in V^{**}$ in the double dual V^{**}.

Show that the mapping $x \to g_x$ is a linear and injective mapping from V to V^{**}, and that $||g_x||^{**} = ||x||$.

If $x \to g_x$ is also surjective, V is said to be *reflexive*.

Show that a Hilbert space is reflexive.

Exercise 77. We consider the space of sequences l^p, where $p \geq 1$. Let $y \in l^q$ where $1/p + 1/q = 1$. (If $p = 1$ then $y \in l^\infty$, the space of bounded sequences.)

Show that

$$x \to \sum_{i=1}^{\infty} x_i \bar{y}_i$$

defines an element $y^* \in (l^p)^*$ with dual norm $||y^*||^* = ||y||_q$.

Exercise 78. Let φ denote a bounded linear functional on a Hilbert space H, and assume that the domain $D(\varphi)$ is a proper subspace of H. Show that there is exactly one extension φ_1 of φ to H with the property that $||\varphi_1|| = ||\varphi||$.

Exercise 79. Let H be a Hilbert space. A mapping $h : H \times H \to \mathbb{C}$ is called *sesquilinear* if, for all $x, x_1, x_2 \in H$ and $\alpha \in \mathbb{C}$, it holds that

$$h(x_1 + x_2, x) = h(x_1, x) + h(x_2, x) \ ,$$
$$h(x, x_1 + x_2) = h(x, x_1) + h(x, x_2) \ ,$$
$$h(\alpha x_1, x_2) = \alpha h(x_1, x_2) \ ,$$
$$h(x_1, \alpha x_2) = \bar{\alpha} h(x_1, x_2) \ .$$

We say that h is *bounded* if there is a constant $c \geq 0$ such that

$$|h(x_1, x_2)| \leq c ||x_1|| \, ||x_2||$$

for all $x_1, x_2 \in H$. The norm $||h||$ is defined as the smallest possible c.

Show that there is a $S \in B(H)$ such that

$$h(x_1, x_2) = (Sx_1, x_2) \ ,$$

and that this representation is unique. Show also that $||h|| = ||S||$.

A sesquilinear form is called *Hermitian* if

$$h(x, y) = \bar{h}(y, x)$$

for all $x, y \in H$. If, moreover, $h(x, x) \geq 0$, the form is called *positive semidefinite*.

Show that in this case we have *Schwarz' inequality:*

$$|h(x, y)|^2 \leq h(x, x)h(y, y)$$

for all $x, y \in H$.

Exercise 80. In the Hilbert space l^2, we define an operator $T : D(T) \rightarrow l^2$ by

$$T((x_n)) = (a_n x_n) ,$$

where (a_n) is a complex sequence.

Find the maximal possible $D(T)$ and show that T is linear. Show that $D(T)$ is dense in l^2.

Show that if (a_n) is bounded, then $D(T) = l^2$ and T is bounded.

Exercise 81. Consider in $L^2(\mathbb{R})$ the operator Q defined by

$$Qf(x) = xf(x) ,$$

with

$$D(Q) = \{f \in L^2(\mathbb{R}) \mid Qf \in L^2(\mathbb{R})\} .$$

Show that Q is linear but not bounded. Show that $D(Q)$ is dense in $L^2(\mathbb{R})$. In quantum mechanics, Q is called the *position operator*.

Exercise 82. Consider in $L^2(\mathbb{R})$ the operator P defined by

$$Pf = -i\frac{\partial f}{\partial x} ,$$

with

$$D(P) = \{f \in L^2(\mathbb{R}) \mid Pf \in L^2(\mathbb{R})\} .$$

Show that P is linear but not bounded. Show that $D(P)$ is dense in $L^2(\mathbb{R})$. In quantum mechanics, P is called the *momentum operator*.

Exercise 83. Let V be a normed space and assume that $T \in B(V)$ is bijective. Show that if T^{-1} is bounded, then

$$\|T^{-1}\| \geq \|T\|^{-1} .$$

Exercise 84. Let (e_k) be an orthonormal basis in a Hilbert space H, and let $T : D(T) \to K$ be a linear operator from H into the Hilbert space K. Show that if $e_k \in D(T)$ for all $k \in \mathbb{N}$, then $D(T)$ is dense in H.

Exercise 85. Let $T : X \to Y$ be a bounded linear operator between two normed spaces, and let $A \subset X$ be compact. Show that $T(A)$ is closed.

Exercise 86. Let T be a self-adjoint operator in a Hilbert space H. Show that if $D(T) = H$, then T is bounded.

Exercise 87. Let T be a bounded linear operator on a Hilbert space H, and assume that M and N are closed subspaces of H.
    Show that

$$T(M) \subset N$$

if and only if

$$T^*(N^\perp) \subset M^\perp .$$

Show, moreover, that

$$\ker(T) = T^*(H)^\perp$$

and

$$\ker(T)^\perp = \overline{T^*(H)}.$$

Exercise 88. Let T be a bounded linear operator on a Hilbert space H with $\|T\| = 1$, and assume that we can find $x_0 \in H$ such that $Tx_0 = x_0$. Show that also $T^*x_0 = x_0$.

Exercise 89. Let (e_n) denote an orthonormal basis in a Hilbert space H, and consider the operator

$$T\left(\sum_{k=1}^{\infty} a_k e_k\right) = \sum_{k=1}^{\infty} a_k e_{k+1} .$$

Find the adjoint T^* and show that T^* is an extension of T^{-1}.

Exercise 90. Let (e_n) denote an orthonormal basis in a Hilbert space H, and consider the operator

$$T(\sum_{k=1}^{\infty} a_k e_k) = \sum_{k=2}^{\infty} \sqrt{k-1} a_k e_{k-1} \ .$$

Show that T is a densely defined, unbounded operator, and find T^*.

Exercise 91. Let $T \in B(H)$. Show that we can write T as

$$T = A + iB$$

where A and B are uniquely determined, bounded, self-adjoint operators.

Exercise 92. Show that $T \in B(H)$ is self-adjoint if and only if one of the following conditions is satisfied:

$$(Tx, x) = (x, Tx) \quad \text{for all} \quad x \in H$$

or

$$(Tx, x) \in \mathbb{R} \quad \text{for all} \quad x \in H.$$

Exercise 93. Let S and T be bounded, self-adjoint operators on a Hilbert space. Show that $ST + TS$ and $i(ST - TS)$ are self-adjoint.

Exercise 94. Let T be a bounded self-adjoint operator. Define the numbers

$$m = \inf\{(Tx, x) \mid \|x\| = 1\}$$

and

$$M = \sup\{(Tx, x) \mid \|x\| = 1\}.$$

Show that $\sigma(T) \subset [m, M]$, and show that both m and M belong to $\sigma(T)$. Show that $\|T\| = \max\{|m|, |M|\}$.

Exercise 95. Consider in $L^2(\mathbb{R})$ the operator Q defined by

$$Qf(x) = xf(x) \ ,$$

with

$$D(Q) = \{f \in L^2(\mathbb{R}) \mid Qf \in L^2(\mathbb{R})\} \ .$$

Show that Q is self-adjoint.

Exercise 96. Show that the set of self-adjoint operators is closed in $B(H)$.

Exercise 97. Let $T \in B(H)$. An operator is *isometric* if $\|Tx\| = \|x\|$ for all $x \in H$. Show that the following conditions are equivalent for $T \in B(H)$:
  a) T is isometric.
  b) $T^*T = I$.
  c) $(Tx, Ty) = (x, y)$ for all $x, y \in H$.

Exercise 98. Let $T \in B(H)$ be an isometric operator. Show that $T(H)$ is a closed subspace. Show that $T(H) = H$ if H is finite dimensional. Give an example of an isometric operator with $T(H) \neq H$.

Exercise 99. Let $T \in B(H)$ be an isometric operator, and let M and N denote closed subspaces of the Hilbert space H. Show that

$$T(M) = N \Rightarrow T(M^\perp) \subseteq N^\perp.$$

Show that T is isometric if and only if, for any orthonormal basis (e_k), the sequence (Te_k) is an orthonormal sequence.

Exercise 100. Let M be a closed linear subspace in the Hilbert space H. Suppose that $T : H \to H$ is a mapping satisfying the following conditions: (i) $(Tx, y) = (x, Ty)$ for all $x, y \in H$; (ii) $Tx = x$ when $x \in M$; and (iii) $Ty = 0$ when $y \in M^\perp$.
  Prove that T is the projection operator with fixed space M.

Exercise 101. Let $T \in B(H)$ be an isometric operator. Show that TT^* is a projection operator and determine the range.

Exercise 102. Consider the Hilbert space $L^2([0, \infty))$. Let $h > 0$ and define the operator T by

$$Tf(x) = 0 \quad \text{for} \quad 0 \leq x < h ,$$
$$Tf(x) = f(x - h) \quad \text{for} \quad h \leq x .$$

Show that T is isometric and determine T^*. Find TT^* and T^*T.

Exercise 103. An operator $T \in B(H)$ is called *unitary* if it is isometric and surjective. Show that the following conditions are equivalent for an operator $T \in B(H)$:
  a) T is unitary.

b) T is bijective and $T^{-1} = T^*$.
c) $T^*T = TT^* = I$.
d) T and T^* are isometric.
e) T is isometric and T^* is injective.
f) T^* is unitary.

Exercise 104. Let (e_k) denote an orthonormal basis in a Hilbert space H, and let $T \in B(H)$ be given by

$$T(\sum_{k=1}^{\infty} a_k e_k) = \sum_{k=1}^{\infty} \lambda_k a_k e_k .$$

Show that T is unitary if and only if $|\lambda_k| = 1$ for all k.

Exercise 105. An operator $T \in B(H)$ is *normal* if

$$TT^* = T^*T.$$

Show that T is normal if and only if $\|T^*x\| = \|Tx\|$ for all $x \in H$.

Exercise 106. An operator $T \in B(H)$ is called *positive* if $(Tx, x) \geq 0$ for all $x \in H$, and we write $T \geq 0$.
Prove the following:
a) $T \geq 0$ implies that T is self-adjoint.
b) If $S, T \geq 0$, $\alpha \geq 0$, then $S + \alpha T \geq 0$.
c) If $T \geq 0$ and $S \in B(H)$, then $S^*TS \geq 0$.
d) If $T \in B(H)$, then $T^*T \geq 0$.
e) If T is an orthogonal projection, then $T \geq 0$.

Exercise 107. Let P_M and P_N denote the orthogonal projections onto the closed subspaces M and N of a Hilbert space H. Show that $M \subset N$ implies that the operator $P_N - P_M$ is positive, and we write $P_M \leq P_N$.

Exercise 108. An operator $T \in B(H)$ is called a *contraction* if

$$\|Tx\| \leq \|x\| \quad \text{for all} \quad x \in H .$$

Show that the following conditions are equivalent for an operator $T \in B(H)$:
a) T is a contraction.
b) $\|T\| \leq 1$.
c) $T^*T \leq I$ ($I - T^*T$ is a positive operator).
d) $TT^* \leq I$ ($I - TT^*$ is a positive operator).

e) T^* is a contraction.

f) T^*T is a contraction.

Exercise 109. Let S and T be linear and bounded operators, and assume that S is compact. Show that ST and TS are compact.

Exercise 110. Let S and T be compact operators in $B(H)$, and let $\alpha \in \mathbb{C}$. Show that $S + \alpha T$ is compact.

Exercise 111. Let T be a bounded operator on a Hilbert space H. Show that:

a) If T is compact, then T^* is also compact.

b) If T^*T is compact, then T is compact.

c) If T is self-adjoint and T^n is compact for some n, then T is compact.

Chapter 5

Exercise 112. Consider the linear operator $T : l^2 \to l^2$ on the space l^2 of absolute square summable complex sequences defined by

$$T(x_1, x_2, \ldots, x_n, \ldots) = \left(\frac{1}{2}x_2, \frac{2}{3}x_3, \ldots, \frac{n}{n+1}x_n, \ldots\right).$$

1) Determine the operator norm $\|T\|$ of T.

2) Find all eigenvalues $\sigma_P(T)$ of T and the corresponding eigenvectors.

3) Determine the resolvent set $\rho(T)$ for T.

4) Determine the adjoint operator T^* and its set of eigenvalues $\sigma_P(T^*)$.

Exercise 113. Consider in $L^2(\mathbb{R})$ the operator Q defined by

$$Qf(x) = xf(x) \, ,$$

with

$$D(Q) = \{f \in L^2(\mathbb{R}) \mid Qf \in L^2(\mathbb{R})\} \, .$$

1) Determine the set of eigenvalues for Q.

2) Determine the resolvent set $\rho(Q)$ for Q.

Exercise 114. Let (e_n) denote an orthonormal basis in a Hilbert space H, and consider the operator

$$T\left(\sum_{k=1}^{\infty} a_k e_k\right) = \sum_{k=1}^{\infty} a_k e_{k+1} \, .$$

Determine $\|T\|$ and $\sigma(T)$.

Exercise 115. Let (e_n) denote an orthonormal basis in a Hilbert space H, and consider the operator

$$T(\sum_{k=1}^{\infty} a_k e_k) = \sum_{k=2}^{\infty} \sqrt{k} a_k e_{k-1} .$$

Determine the spectrum $\sigma(T)$, and find for each eigenvalue the corresponding eigenvectors.

Exercise 116. Let (e_n) denote an orthonormal basis in a Hilbert space H. We define the sequence $(f_k)_{k\in\mathbb{Z}}$ by

$$f_k = \begin{cases} e_1 & \text{for} \quad k = 0 \\ e_{2k+1} & \text{for} \quad k > 0 \\ e_{-2k} & \text{for} \quad k < 0 . \end{cases}$$

In this way $(f_k)_{k\in\mathbb{Z}}$ is an orthonormal basis. We define the *double-sided shift operator* S by

$$S(\sum_{k=-\infty}^{\infty} a_k f_k) = \sum_{k=-\infty}^{\infty} a_k f_{k+1} .$$

Show that S is a bounded operator, and show that S has no eigenvalues.

Exercise 117. For a real number $h \neq 0$, define the operator τ_h on $L^2(\mathbb{R})$ by

$$\tau_h f(x) = f(x - h) \quad \text{for } x \in \mathbb{R}.$$

1) Show that τ_h has no eigenvalues.
2) Show that all complex numbers λ with $|\lambda| \neq 1$ belong to the resolvent set $\rho(\tau_h)$.
(It is, in fact, true that $\sigma(\tau_h) = \{\lambda \in \mathbb{C} \mid |\lambda| = 1\}$.)

Exercise 118. Let $T \in B(H)$ where H is a Hilbert (or just Banach) space. Show that $\|R_\lambda(T)\| \to 0$ for $|\lambda| \to \infty$.

Exercise 119. Let (e_n) denote an orthonormal basis in a Hilbert space H, and let (r_k) be all the rational numbers in $]0,1[$ arranged as a sequence. Consider the operator

$$T(\sum_{k=1}^{\infty} a_k e_k) = \sum_{k=1}^{\infty} r_k a_k e_k .$$

Show that T is self-adjoint and that $\|T\| = 1$. Find $\rho(T)$ and determine the point spectrum and the continuous spectrum for T.

Exercise 120. Let $T \in B(H)$ be unitary. Show that

$$\sigma(T) \subset \{z \in \mathbb{C} \mid |z| = 1\} .$$

Exercise 121. Let $T \in B(H)$ be normal. Show that

$$\|(T - \lambda I)x\| = \|(T^* - \bar{\lambda}I)x\|$$

for all $x \in H$. Show that $\sigma_r(T)$ is empty.

Exercise 122. Let (e_k) denote an orthonormal basis in a Hilbert space H, and define the operator T by

$$T(\sum_{k=1}^{\infty} a_k e_k) = \sum_{k=2}^{\infty} \frac{1}{k} a_k e_{k-1} .$$

Show that T is compact and find T^*. Find $\sigma_p(T)$ and $\sigma_p(T^*)$.

Exercise 123. Let (e_k) denote an orthonormal basis in a Hilbert space H, and assume that the operator T has the matrix representation (t_{jk}) with respect to the basis (e_k). Show that

$$\sum_{j=1}^{\infty} \sum_{k=1}^{\infty} |t_{jk}|^2 < \infty$$

implies that T is compact.

Let (f_k) denote another orthonormal basis in H, and let $s_{jk} = (Tf_j, f_k)$ so that (s_{jk}) is the matrix representation of T with respect to the basis (f_k). Show that

$$\sum_{j=1}^{\infty} \sum_{k=1}^{\infty} |t_{jk}|^2 = \sum_{j=1}^{\infty} \sum_{k=1}^{\infty} |s_{jk}|^2 .$$

An operator satisfying $\sum_{j=1}^{\infty} \sum_{k=1}^{\infty} |t_{jk}|^2 < \infty$ is called a *general Hilbert-Schmidt operator*.

Exercise 124. For a general Hilbert-Schmidt operator, we define the Hilbert-Schmidt norm $\|\cdot\|_{HS}$ by

$$\|T\|_{HS} = (\sum_{j=1}^{\infty}\sum_{k=1}^{\infty}|t_{jk}|^2)^{\frac{1}{2}} .$$

Show that this *is* a norm, and show that

$$\|T\| \le \|T\|_{HS}$$

for a general Hilbert-Schmidt operator T.

Exercise 125. Define for $f \in L^2(\mathbb{R})$ the operator K by

$$Kf(x) = \int_{-\infty}^{\infty} \frac{1}{2}e^{-|x-t|}f(t)dt .$$

Show that $Kf \in L^2(\mathbb{R})$, and that K is linear and bounded with norm ≤ 1.
Show that the function $\frac{1}{2}e^{-|x-t|}$ does not belong to $L^2(\mathbb{R}^2)$, so that K is not a Hilbert-Schmidt operator.

Chapter 6

Exercise 126. Let X be an arbitrary vector space, and let $S,T : X \to X$ be linear operators satisfying $ST = TS$ (commuting operators). Suppose further, that ST has an inverse operator $(ST)^{-1} : X \to X$.
Prove that the linear operators S and T both have inverse operators, respectively $S^{-1} = T(ST)^{-1}$ and $T^{-1} = S(ST)^{-1}$.

Exercise 127. Let $T : X \to Y$ be a bounded linear operator between Hilbert spaces X and Y. Denote by $T_a^* : Y \to X$ the adjoint operator of T, and by $T_d^* : Y^* \to X^*$ the dual operator of T.
Let $R_X : X \to X^*$ and $R_Y : Y \to Y^*$ be the equivalences of normed vector spaces defined by the Riesz' Representation Theorem 3.5.10. Precisely, $(R_X(z))(x) = (x, z)_X$ for $x, z \in X$, and $(R_Y(z))(y) = (y, z)_Y$ for $y, z \in Y$.
Prove that $T_d^* R_Y = R_X T_a^*$, implying that $T_d^* = R_X T_a^* R_Y^{-1}$.

Exercise 128. Let X be an arbitrary Banach space with norm $\|\cdot\|$.
Prove that the dual space X^* of all bounded linear functionals $\phi : X \to \mathbb{K}$, is a Banach space when equipped with the operator norm

$$\|\phi\| = \sup_{x \in X} \{|\phi(x)| \mid \|x\| = 1\} .$$

Exercise 129. Let $P : X \to X$ be a compact, idempotent (i.e. $P^2 = P$) linear operator on the Banach space X. Show that P has finite rank.

Exercise 130. Let $T : X \to Y$ be a compact linear operator between Banach spaces. If $T(X)$ is a closed subspace in Y, show that it has finite dimension.

Exercise 131. Let M be a closed subspace and N a finite dimensional subspace of the normed vector space X. Suppose $M \cap N = \{0\}$, and consider the direct sum $M \oplus N$ of M and N. Prove that $M \oplus N$ is a closed linear subspace of X.
Hint: You can exploit that closed balls in N are compact.

Exercise 132. Let T be a bounded linear operator on a Banach space X.
Suppose there exists a constant $c \in \mathbb{R}$ such that

$$||x|| \leq c||Tx|| \quad \text{for all } x \in X.$$

Prove that the image $T(X)$ of T is a closed linear subspace of X.

Exercise 133. Let $T_1 : X_1 \to Y_1$ and $T_2 : X_2 \to Y_2$ be bounded linear operators between Banach spaces.
1) Define the linear mapping $T_1 \otimes T_2 : X_1 \otimes X_2 \to Y_1 \otimes Y_2$ between product vector spaces by

$$T_1 \otimes T_2(x_1, x_2) = (T_1(x_1), T_2(x_2)) \quad \text{for} \quad (x_1, x_2) \in X_1 \otimes X_2.$$

Show that $T_1 \otimes T_2$ is a bounded linear operator with operator norm

$$||T_1 \otimes T_2|| = ||T_1|| + ||T_2||.$$

2) Show that if T_1 and T_2 are Fredholm operators, then $T_1 \otimes T_2$ is a Fredholm operator and

$$\text{index}\,(T_1 \otimes T_2) = \text{index}\,(T_1) + \text{index}\,(T_2).$$

Exercise 134. Consider the Banach spaces l^p, $p \geq 1$, of absolute p-summable sequences. Let $\mathcal{F}(l^p)$ be the set of Fredholm operators on l^p, and equip $\mathcal{F}(l^p)$ with the product structure defined by composition of operators.
Show that index : $\mathcal{F}(l^p) \to \mathbb{Z}$ defines a surjective homomorphism of $\mathcal{F}(l^p)$ (with product structure) onto the integers $\mathbb{Z}$ (with additive structure).
Show that the result holds more generally for all Banach spaces X which are separable by a basis.

Exercise 135. Let T be an arbitrary non-trivial bounded linear operator on a Banach space X. Show that the Fredholm equation $T(f) - \lambda f = g$ on X has a unique solution $f \in X$ for every $g \in X$, if $0 < \|T\| < |\lambda|$.

Exercise 136. Consider the integral operator T on the Banach space of continuous functions $C([0, 2\pi])$ determined by the continuous kernel

$$\Phi(x, y) = \sum_{k=0}^{\infty} \frac{1}{(k+1)^2} \Big(\cos\big((k+1)x\big) \sin\big(ky\big) - \sin\big((k+1)x\big) \cos\big(ky\big) \Big)$$

for $x, y \in [0, 2\pi]$.

The operator $T : C([0, 2\pi]) \to C([0, 2\pi])$ is then defined by

$$T(f)(y) = \int_0^{2\pi} \Phi(x, y) f(x) dx \quad \text{for} \quad f \in C([0, 2\pi]), \ y \in [0, 2\pi].$$

Show that T has no eigenvalues.

Exercise 137. Let T denote the Hilbert-Schmidt operator on $L^2[0, 2\pi]$ with continuous kernel

$$\Phi(x, y) = \sin(x) \cos(y) \quad \text{for} \quad 0 \le x, y \le 2\pi .$$

Show that 0 is the only eigenvalue for T.
Find an orthonormal basis for $\ker(T)$.

Exercise 138. Let T denote the Hilbert-Schmidt operator on $L^2[0, 1]$ with continuous kernel

$$\Phi(x, y) = x + y \quad \text{for} \quad 0 \le x, y \le 1 .$$

Find all eigenvalues and eigenfunctions for T.
Solve the equation $Tf = \mu f + g$, for a given $g \in L^2([0, 1])$, when μ is not in the spectrum for T.

Exercise 139. Let T denote the Hilbert-Schmidt operator on $L^2([-\frac{\pi}{2}, \frac{\pi}{2}])$ with continuous kernel

$$\Phi(x, y) = \cos(x - y) \quad \text{for} \quad -\frac{\pi}{2} \le x, y \le \frac{\pi}{2} .$$

Find all eigenvalues and eigenfunctions for T.
Solve the equation $Tf = \mu f + g$, for a given $g \in L^2([-\frac{\pi}{2}, \frac{\pi}{2}])$, when μ is not in the spectrum for T.

Exercise 140. Let T denote the Hilbert-Schmidt operator on $L^2([-\pi, \pi])$ with continuous kernel

$$\Phi(x, y) = \left(\cos(x) + \cos(y) \right)^2 \quad \text{for} \quad -\pi \leq x, y \leq \pi \ .$$

Find all eigenvalues and eigenfunctions for T, and find an orthonormal basis for $\ker(T)$.

Exercise 141. Let T denote a self-adjoint Hilbert-Schmidt operator on $L^2([0, 1])$ with continuous kernel Φ. Denote by $\|T\|$ and $\|\Phi\|_2$ respectively the operator norm of T and the 2-norm of Φ.

Show that $\|T\| = \|\Phi\|_2$ if and only if the spectrum for T consists of only two points.

Bibliography

Main sources:

V.L. Hansen: *Fundamental Concepts in Modern Analysis*, World Scientific, 1999.

M. Pedersen: *Functional Analysis in Applied Mathematics and Engineering*, Chapman & Hall/CRC, 2000.

Other sources:

S.K. Berberian: *Introduction to Hilbert Space*, Oxford University Press, 1961.

Y. Choquet-Bruhat, C. de Witt-Morette, M. Dillard-Bleick: *Analysis, Manifolds and Physics*, North-Holland Publishing Company, 1977.

D. Gilbarg, N.S. Trudinger: *Elliptic Partial Differential Equations of Second Order*, Springer-Verlag, 2nd ed. 1983.

L. Hörmander: *The Analysis of Linear Partial Differential Operators III*, Springer-Verlag, 1985.

R. Kress: *Linear Integral Equations*, Applied Mathematical Sciences 82, Springer-Verlag, 2014.

M.E. Munroe: *Introduction to Measure and Integration*, Addison-Wesley Publishing Company Inc., 1953.

R.S. Palais: *Seminar on the Atiyah-Singer Index Theorem*, Annals of Mathematics Studies 57, Princeton University Press, 1965.

W. Rudin: *Functional Analysis*, McGraw-Hill Inc., 1973.

K. Yosida: *Functional Analysis*, Springer-Verlag, 1965.

Suggestions for further reading:

D.D. Bleecker, B. Booss-Bavnbek: *Index Theory with Applications to Mathematics and Physics*, International Press of Boston, Inc., 2013.

J.B. Conway: *A Course in Functional Analysis*, Springer-Verlag, 2nd ed. 1990.

P.D. Lax: *Functional Analysis*, John Wiley & Sons Inc., 2002.

G.K. Pedersen: *Analysis Now*, Springer-Verlag, 1989.

W. Rudin: *Real and Complex Analysis*, McGraw-Hill Book Co., 1966.

List of Symbols

Symbol	Explanation	Page
$\ker(\phi)$	kernel of linear functional ϕ	77
$\mu(A)$	Lebesgue measure of set A	45
L^p	space of pth power Lebesgue integrable functions	42
l^p	the space of absolute p-summable sequences	51
$M^\perp$	the orthogonal complement to subset M	74
$V_1 \oplus V_2$	direct sum of vector spaces	115
$V \otimes W$	product vector space	23
$\|\cdot\|_p$	p-norm	38
P	projection operator	88
$[V/U]$	quotient space of vector space V modulo subspace U	113
$\rho(T)$	resolvent set for linear operator T	96
$\mathcal{R}_0(\mathbb{R})$	the space of Riemann integrable functions	49
$\sigma(T)$	spectrum for linear operator T	96
$\sigma_C(T)$	continuous spectrum for linear operator T	97
$\sigma_P(T)$	point spectrum for linear operator T	97
$\sigma_r(T)$	residual spectrum for linear operator T	97
$\mathcal{B}(X,Y)$	space of bounded operators between Banach spaces	126
$\mathcal{C}(X,Y)$	space of compact operators between Banach spaces	126
$\mathcal{F}(X,Y)$	space of Fredholm operators between Banach spaces	126
$\mathcal{I}(X,Y)$	space of invertible operators between Banach spaces	126
$\sum_{n=1}^{\infty} x_n$	sum of series	61
$\mathrm{support}(f)$	support of function	38
$\|\cdot\|_\infty$	uniform norm	29
$x_n \rightharpoonup x$	weak convergence of sequence (x_n) to x	77

Index

Printed in the United States
By Bookmasters